Developing Credit Risk Models Using SAS® Enterprise Miner™ and SAS/STAT®

Theory and Applications

Iain L. J. Brown, PhD

support.sas.com/bookstore

Contents

About This Book

Purpose

This book sets out to empower readers with both theoretical and practical skills for developing credit risk models for Probability of Default (PD), Loss Given Default (LGD) and Exposure At Default (EAD) models using SAS Enterprise Miner and SAS/STAT. From data pre-processing and sampling, through segmentation analysis and model building and onto reporting and validation, this text aims to explain through theory and application how credit risk problems are formulated and solved.

Is This Book for You?

Those who will benefit most from this book are practitioners (particularly analysts) and students wishing to develop their statistical and industry knowledge of the techniques required for modelling credit risk parameters. The step-by-step guide shows how models can be constructed through the use of SAS technology and demonstrates a best-practice approach to ensure accurate and timely decisions are made. Tutorials at the end of the book detail how to create projects in SAS Enterprise Miner and walk through a typical credit risk model building process.

Prerequisites

In order to make the most of this text, a familiarity with statistical modelling is beneficial. This book also assumes a foundation level of SAS programming skills. Knowledge of SAS Enterprise Miner is not required, as detailed use cases will be given.

Scope of This Book

This book covers the use of SAS statistical programming (Base SAS, SAS/STAT, SAS Enterprise Guide), SAS Enterprise Miner in the development of credit risk models, and a small amount of SAS Model Manager for model monitoring and reporting.

This book does not provide proof of the statistical algorithms used. References and further readings to sources where readers can gain more information on these algorithms are given throughout this book.

About the Examples

Software Used to Develop the Book's Content

SAS 9.4

SAS/STAT 12.3

SAS Enterprise Guide 6.1

SAS Enterprise Miner 12.3 (with Credit Scoring nodes)

SAS Model Manager 12.3

Example Code and Data

You can access the example code and data for this book by linking to its author page at http://support.sas.com/publishing/authors. Select the name of the author. Then, look for the cover thumbnail of this book, and select Example Code and Data to display the SAS programs that are included in this book.

For an alphabetical listing of all books for which example code and data is available, see http://support.sas.com/bookcode. Select a title to display the book's example code.

If you are unable to access the code through the website, send e-mail to saspress@sas.com.

Additional Resources

SAS offers you a rich variety of resources to help build your SAS skills and explore and apply the full power of SAS software. Whether you are in a professional or academic setting, we have learning products that can help you maximize your investment in SAS.

Bookstore	http://support.sas.com/bookstore/
Training	http://support.sas.com/training/
Certification	http://support.sas.com/certify/
SAS Global Academic Program	http://support.sas.com/learn/ap/
SAS OnDemand	http://support.sas.com/learn/ondemand/
Support	http://support.sas.com/techsup/
Training and Bookstore	http://support.sas.com/learn/
Community	http://support.sas.com/community/

Keep in Touch

We look forward to hearing from you. We invite questions, comments, and concerns. If you want to contact us about a specific book, please include the book title in your correspondence.

To Contact the Author through SAS Press

By e-mail: saspress@sas.com

Via the Web: http://support.sas.com/author_feedback

SAS Books

For a complete list of books available through SAS, visit http://support.sas.com/bookstore. Phone: 1-800-727-0025

Fax: 1-919-677-8166

E-mail: sasbook@sas.com

SAS Book Report

Receive up-to-date information about all new SAS publications via e-mail by subscribing to the SAS Book Report monthly eNewsletter. Visit http://support.sas.com/sbr.

Publish with SAS

SAS is recruiting authors! Are you interested in writing a book? Visit http://support.sas.com/saspress for more information.

Data Mining with SAS Enterprise Miner

SAS Enterprise Miner streamlines the data mining process to create highly accurate predictive and descriptive models based on analysis of vast amounts of data from across an enterprise. Data mining is applicable in a variety of industries and provides methodologies for such diverse business problems as fraud detection, customer retention and attrition, database marketing, market segmentation, risk analysis, affinity analysis, customer satisfaction, bankruptcy prediction, and portfolio analysis.

In SAS Enterprise Miner, the data mining process has the following (SEMMA) steps:

- Sample the data by creating one or more data sets. The sample should be large enough to contain significant information, yet small enough to process. This step includes the use of data preparation tools for data importing, merging, appending, and filtering, as well as statistical sampling techniques.

- Explore the data by searching for relationships, trends, and anomalies in order to gain understanding and ideas. This step includes the use of tools for statistical reporting and graphical exploration, variable selection methods, and variable clustering.

- Modify the data by creating, selecting, and transforming the variables to focus the model selection process. This step includes the use of tools for defining transformations, missing value handling, value recoding, and interactive binning.

- Model the data by using the analytical tools to train a statistical or machine learning model to reliably predict a desired outcome. This step includes the use of techniques such as linear and logistic regression, decision trees, neural networks, partial least squares, LARS and LASSO, nearest neighbor, and importing models defined by other users or even outside SAS Enterprise Miner.

- Assess the data by evaluating the usefulness and reliability of the findings from the data mining process. This step includes the use of tools for comparing models and computing new fit statistics, cutoff analysis, decision support, report generation, and score code management.

You might or might not include all of the SEMMA steps in an analysis, and it might be necessary to repeat one or more of the steps several times before you are satisfied with the results.

After you have completed the SEMMA steps, you can apply a scoring formula from one or more champion models to new data that might or might not contain the target variable. Scoring new data that is not available at the time of model training is the goal of most data mining problems.

Furthermore, advanced visualization tools enable you to quickly and easily examine large amounts of data in multidimensional histograms and to graphically compare modeling results.

Scoring new data that is not available at the time of model training is the goal of most data mining exercises. SAS Enterprise Miner includes tools for generating and testing complete score code for the entire process flow diagram as SAS Code, C code, and Java code, as well as tools for interactively scoring new data and examining the results. You can register your model to a SAS Metadata Server to share your results with users of applications such as SAS Enterprise Guide that can integrate the score code into reporting and production processes. SAS Model Manager complements the data mining process by providing a structure for managing projects through development, testing, and production environments and is fully integrated with SAS Enterprise Miner.

About Credit Scoring for SAS Enterprise Miner

The additional add-in of credit scoring for SAS Enterprise Miner facilitates analysts in building, validating, and deploying credit risk models. It enables organizations to create credit scorecards using in-house expertise and resources to decide whether to accept an applicant (application scoring): to determine the likelihood of defaults among customers who have already been accepted (behavioral scoring); and to predict the likely amount of debt that the lender can expect to recover (collection scoring). Three key additional nodes are made available to analysts: the Interactive Grouping node, Scorecard node and Reject Inference node. **Note:** The Credit Scoring for SAS Enterprise Miner solution is not included with the base version of SAS Enterprise Miner. If your site has not licensed Credit Scoring for SAS Enterprise Miner, the credit scoring node tools do not appear in your SAS Enterprise Miner software.

SAS Enterprise Miner and the Credit Scoring nodes will be used extensively throughout this book to demonstrate the data exploration and modelling processes discussed. A tutorial section is also located at the end of this book which gives a step-by-step walkthrough of typical tasks using SAS Enterprise Miner.

About SAS/STAT

SAS/STAT software provides comprehensive statistical tools for a wide range of statistical analyses, including analysis of variance, categorical data analysis, cluster analysis, multiple imputation, multivariate analysis, nonparametric analysis, power and sample size computations, psychometric analysis, regression, survey data analysis, and survival analysis. A few examples include nonlinear mixed models, generalized linear models, correspondence analysis, and robust regression. The software is constantly being updated to reflect new methodology. In addition to more than 80 procedures for statistical analysis, SAS/STAT software also includes the Power and Sample Size Application (PSS), an interface to power and sample size computations.

SAS/STAT code is used extensively throughout this book to demonstrate data manipulation and model development coded tasks.

About The Author

Dr. Iain Brown is an Analytics Specialist Consultant at SAS, specializing in Credit Risk. Prior to joining SAS in 2011, he worked as a Credit Risk Analyst at a major UK retail bank where he built and validated PD, LGD, and EAD models using SAS software. He has spoken at a number of internationally renowned conferences and conventions and has published papers on the topic of credit risk modeling in the International Journal of Forecasting and the Journal of Expert Systems with Applications. In 2011, he won the SAS Student Ambassador award for his doctoral research, which recognizes and supports students who use SAS technologies in innovative ways to benefit their respective industries and fields of study.

Iain has a BBA in Business from the University of Kent, an MSc in Operational Research from the London School of Economics and Political Science (LSE), and a PhD in Credit Risk from the University of Southampton. Iain is also an active member of the Operational Research (OR) Society; in July 2014, he was awarded the title of Associate Fellow of the OR Society (AFORS) for his contribution to the field of OR. His research interests include data mining, credit scoring, credit risk modeling, and Basel compliancy.

Learn more about this author by visiting his author page at http://support.sas.com/publishing/authors/brown_iain.html. There, you can download free book excerpts, access example code and data, read the latest reviews, get updates, and more.

Acknowledgments

This book would not have been possible without the support and guidance of a number of important people whom I would like to take this opportunity to acknowledge and thank.

Among the many people I would like to thank are all those people at SAS involved in making this book possible, including my acquisition editor Shelley Sessoms, technical editors Naeem Siddiqi and Jim Seabolt, whose input has greatly enhanced the text, and a special thanks to Stephenie Joyner and Brenna Leath, my developmental editors, for keeping me on track. I am also greatly thankful to the SAS UK Analytics Practice, in particular Dr. Laurie Miles, John Spooner, and Colin Gray, for their support and guidance throughout my career at SAS.

I would also like to pay a special thanks to my supervising team during my doctoral research, Dr, Christophe Mues, Prof. Lyn Thomas, and Dr. Bart Baesens. Without their joint tutorage and expert knowledge in the field of credit risk modelling, I could not have achieved much of the work conducted in this book. I have also had the great pleasure of working alongside a number of well-established and flourishing academics in the field of credit risk and credit scoring. I would like to thank the team I worked alongside on the LGD benchmarking case study referenced in Chapter 4: Dr. Gert Loterman, Dr. David Martens, Dr. Bart Baesens, and Dr. Christophe Mues.

Finally, I would like to express my utmost thanks to my wife, parents, sister, and whole family, without whose support I would not have achieved any of the goals I have set out to attain.

Chapter 1 Introduction

1.1 Book Overview

This book aims to define the concepts underpinning credit risk modeling and to show how these concepts can be formulated with practical examples using SAS software. Each chapter tackles a different problem encountered by practitioners working or looking to work in the field of credit risk and give a step-by-step approach to leverage the power of the SAS Analytics suite of software to solve these issues.

This chapter begins by giving an overview of what credit risk modeling entails, explaining the concepts and terms that one would typically come across working in this area. We then go on to scrutinize the current regulatory environment, highlighting the key reporting parameters that need to be estimated by financial institutions subject to the Basel capital requirements. Finally, we discuss the SAS analytics software used for the analysis part of this book.

The remaining chapters are structured as follows:

Chapter 2 covers the area of sampling and data pre-processing. This chapter defines and contextualizes issues such as variable selection, missing values, and outlier detection within the area of credit risk modeling, and gives practical applications of how these issues can be solved.

Chapter 3 details the theory and practical aspects behind the creation of Probability of Default (PD) models. This focuses on standard and novel modeling techniques, shows how each of these can be used in the estimation of PD, and demonstrates the full development of an application and behavioral scorecard using SAS Enterprise Miner.

Chapter 4 focuses on the development of Loss Given Default (LGD) models and the considerations with regard to the distribution of LGD that have to be made for modeling this parameter. A variety of modeling approaches are discussed and compared in a case study in order to show how improvements over the traditional industry approach of linear regression can be made.

Chapter 5 defines the concept of Exposure at Default (EAD) and how this parameter is formulated and estimated. A full model development process is shown through practical examples. The aim of this chapter is to fully explore the implications of model choice, input variables, and how best to estimate EAD.

Chapter 6 defines and explains the concepts of stress testing under the three pillars of the Basel Capital Accord and what this entails for financial institutions.

Chapter 7 focuses on how model reports can be generated from the procedures and methodologies created throughout this book. This chapter covers the key reporting outputs required within the regulatory framework and shows through SAS Model Manager and example code how these outputs can be created.

By the conclusion of this book, readers will have a comprehensive guide to developing credit risk models both from a theoretical and practical perspective. We also aim to show how analysts can create and implement credit risk models using example code and projects in SAS.

1.2 Overview of Credit Risk Modeling

With cyclical financial instabilities in the credit markets, the area of credit risk modeling has become ever more important, leading to the need for more accurate and robust models. Since the introduction of the Basel II Capital Accord (Basel Committee on Banking Supervision, 2004) over a decade ago, qualifying financial institutions have been able to derive their own internal credit risk models under the advanced internal ratings-based approach (A-IRB) without relying on regulator's fixed estimates.

The Basel II Capital Accord prescribes the minimum amount of regulatory capital an institution must hold so as to provide a safety cushion against unexpected losses. Under the advanced internal ratings-based approach (A-IRB), the accord allows financial institutions to build risk models for three key risk parameters: Probability of Default (PD), Loss Given Default (LGD), and Exposure at Default (EAD). PD is defined as the likelihood that a loan will not be repaid and will therefore fall into default. LGD is the estimated economic loss, expressed as a percentage of exposure, which will be incurred if an obligor goes into default. EAD is a measure of the monetary exposure should an obligor go into default. These topics will be explained in more detail in the next section.

With the arrival of Basel III and as a response to the latest financial crisis, the objective to strengthen global capital standards has been reinstated. A key focus here is the reduction in reliance on external ratings by the financial institutions, as well as a greater focus on stress testing. Although changes are inevitable, a key point worth noting is that with Basel III there is no major impact on underlying credit risk models. Hence the significance in creating these robust risk models continues to be of paramount importance.

In this book, we use theory and practical applications to show how these underlying credit risk models can be constructed and implemented through the use of SAS (in particular, SAS Enterprise Miner and SAS/STAT). To achieve this, we present a comprehensive guide to the classification and regression techniques needed to develop models for the prediction of all three components of expected loss: PD, LGD and EAD. The reason why these particular topics have been chosen is due in part to the increased scrutiny on the financial sector and the pressure placed on them by the financial regulators to move to the advanced internal ratings-based approach. The financial sector is therefore looking for the best possible models to determine their minimum capital requirements through the estimation of PD, LGD and EAD.

This introduction chapter is structured as follows. In the next section, we give an overview of the current regulatory environment, with emphasis on its implications to credit risk modeling. In this section, we explain the three key components of the minimum capital requirements: PD, LGD and EAD. Finally, we discuss the SAS software used in this book to support the practical applications of the concepts covered.

1.3 Regulatory Environment

The banking/financial sector is one of the most closely scrutinized and regulated industries and, as such, is subject to stringent controls. The reason for this is that banks can only lend out money in the form of loans if depositors trust that the bank and the banking system is stable enough and their money will be there when they require to withdraw it. However, in order for the banking sector to provide personal loans, credit cards, and mortgages, they must leverage depositors' savings, meaning that only with this trust can they continue to function. It is imperative, therefore, to prevent a loss of confidence and distrust in the banking sector from occurring, as it can have serious implications to the wider economy as a whole.

The job of the regulatory bodies is to contribute to ensuring the necessary trust and stability by limiting the level of risk that banks are allowed to take. In order for this to work effectively, the maximum risk level banks can take needs to be set in relation to the bank's own capital. From the bank's perspective, the high cost of acquiring and holding capital makes it prohibitive and unfeasible to have it fully cover all of a bank's risks. As a compromise, the major regulatory body of the banking industry, the Basel Committee on Banking Supervision, proposed guidelines in 1988 whereby a solvability coefficient of eight percent was introduced. In other words, the total assets, weighted for their risk, must not exceed eight percent of the bank's own capital (SAS Institute, 2002).

The figure of eight percent assigned by the Basel Committee was somewhat arbitrary, and as such, this has been subject to much debate since the conception of the idea. After the introduction of the Basel I Accord, more than one hundred countries worldwide adopted the guidelines, marking a major milestone in the history of global banking regulation. However, a number of the accord's inadequacies, in particular with regard to the way that credit risk was measured, became apparent over time (SAS Institute, 2002). To account for these issues, a revised accord, Basel II, was conceived. The aim of the Basel II Capital Accord was to further strengthen the financial sector through a three pillar approach. The following sections detail the current state of the regulatory environment and the constraints put upon financial institutions.

1.3.1 Minimum Capital Requirements

The Basel Capital Accord (Basel Committee on Banking Supervision, 2001a) prescribes the minimum amount of regulatory capital an institution must hold so as to provide a safety cushion against unexpected losses. The Accord is comprised of three pillars, as illustrated by Figure 1.1:

Pillar 1: Minimum Capital Requirements

Pillar 2: Supervisory Review Process

Pillar 3: Market Discipline (and Public Disclosure)

Figure 1.1: Pillars of the Basel Capital Accord

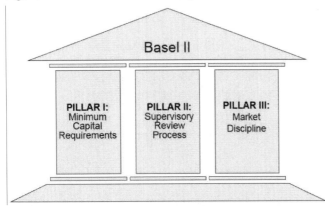

Pillar 1 aligns the minimum capital requirements to a bank's actual risk of economic loss. Various approaches to calculating this are prescribed in the Accord (including more risk-sensitive standardized and internal ratings-based approaches) which will be described in more detail and are of the main focus of this text. Pillar 2 refers to supervisors evaluating the activities and risk profiles of banks to determine whether they should hold higher levels of capital than those prescribed by Pillar 1, and offers guidelines for the supervisory review process, including the approval of internal rating systems. Pillar 3 leverages the ability of market discipline to motivate prudent management by enhancing the degree of transparency in banks' public disclosure (Basel, 2004).

Pillar 1 of the Basel II Capital Accord entitles banks to compute their credit risk capital in either of two ways:

1. Standardized Approach
2. Internal Ratings-Based (IRB) Approach
   a. Foundation Approach
   b. Advanced Approach

Under the standardized approach, banks are required to use ratings from external credit rating agencies to quantify required capital. The main purpose and strategy of the Basel committee is to offer capital incentives to banks that move from a supervisory approach to a best-practice advanced internal ratings-based approach. The two versions of the internal ratings-based (IRB) approach permit banks to develop and use their own internal risk ratings, to varying degrees. The IRB approach is based on the following four key parameters:

1. Probability of Default (PD): the likelihood that a loan will not be repaid and will therefore fall into default in the next 12 months;
2. Loss Given Default (LGD): the estimated economic loss, expressed as a percentage of exposure, which will be incurred if an obligor goes into default - in other words, LGD equals: 1 minus the recovery rate;
3. Exposure At Default (EAD): a measure of the monetary exposure should an obligor go into default;
4. Maturity (M): is the length of time to the final payment date of a loan or other financial instrument.

The internal ratings-based approach requires financial institutions to estimate values for PD, LGD, and EAD for their various portfolios. Two IRB options are available to financial institutions: a foundation approach and an advanced approach (Figure 1.2) (Basel Committee on Banking Supervision, 2001a).

Figure 1.2: Illustration of Foundation and Advanced Internal Ratings-Based (IRB) approach

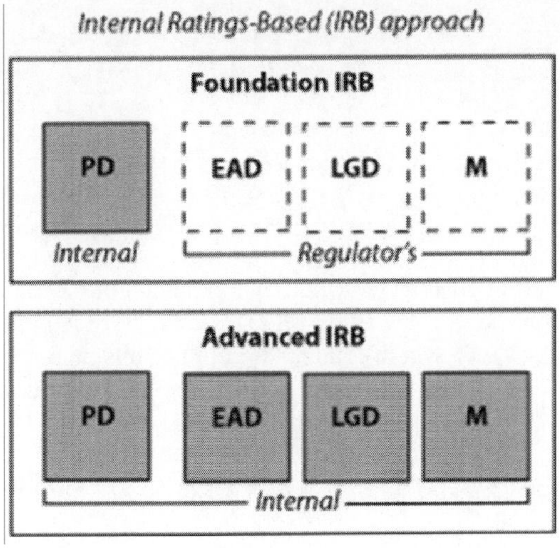

The difference between these two approaches is the degree to which the four parameters can be measured internally. For the foundation approach, only PD may be calculated internally, subject to supervisory review (Pillar 2). The values for LGD and EAD are fixed and based on supervisory values. For the final parameter, M, a single average maturity of 2.5 years is assumed for the portfolio. In the advanced IRB approach, all four parameters are to be calculated by the bank and are subject to supervisory review (Schuermann, 2004).

Under the A-IRB, financial institutions are also recommended to estimate a "Downturn LGD", which 'cannot be less than the long-run default-weighted average LGD calculated based on the average economic loss of all observed defaults with the data source for that type of facility' (Basel, 2004).

1.3.2 Expected Loss

Financial institutions expect a certain number of the loans they make to go into default; however they cannot identify in advance which loans will default. To account for this risk, a value for expected loss is priced into the products they offer. Expected Loss (EL) can be defined as the expected means loss over a 12 month period from which a basic premium rate is formulated. Regulatory controllers assume organizations will cover EL through loan loss provisions. Consumers experience this provisioning of expected loss in the form of the interest rates organizations charge on their loan products.

To calculate this value, the PD of an entity is multiplied by the estimated LGD and the current exposure if the entity were to go into default.

From the parameters, PD, LGD and EAD, expected loss (EL) can be derived as follows:

$$EL = PD \times LGD \times EAD \quad \text{(1.1)}$$

For example, if PD = 2%, LGD = 40% and EAD = $10,000, then EL would equal $80. Expected Loss can also be measured as a percentage of EAD:

$$EL\% = PD \times LGD \quad \text{(1.2)}$$

In the previous example, expected loss as a percentage of EAD would be equal to $EL\% = 0.8\%$.

1.3.3 Unexpected Loss

Unexpected loss is defined as any loss on a financial product that was not expected by a financial organization and therefore not factored into the price of the product. The purpose of the Basel regulations is to force banks to retain capital to cover the entire amount of the Value-at-Risk (VaR), which is a combination of this unexpected loss plus the expected loss. Figure 1.3 highlights the Unexpected Loss, where UL is the difference between the Expected Loss and a 1 in 1000 chance level of loss.

Figure 1.3: Illustration of the Difference between Expected/Unexpected Loss and a 1 in 1000 Chance Level of Loss

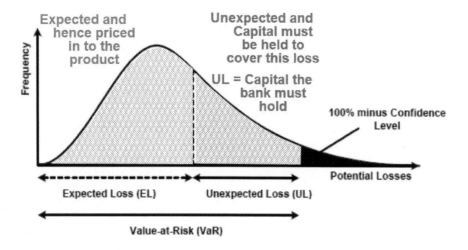

1.3.4 Risk Weighted Assets

Risk Weighted Assets (RWA) are the assets of the bank (money lent out to customers and businesses in the form of loans) accounted for by their riskiness. The RWA are a function of PD, LGD, EAD and M, where K is the capital requirement:

$$RWA = (12.5) \times K \times EAD \quad \text{(1.3)}$$

Under the Basel capital regulations, all banks must declare their RWA, hence the importance in estimating the three components, PD, LGD, and EAD, which go towards the formulation of RWA. The multiplication of the capital requirement (K) by 12.5 $\left(\dfrac{1}{12.5} = 0.08 \right)$ is to ensure capital is no less than 8% of RWA. Figure 1.4 is a graphical representation of RWA and shows how each component feeds into the final RWA value.

Figure 1.4: Relationship between PD, LGD, EAD and RWA

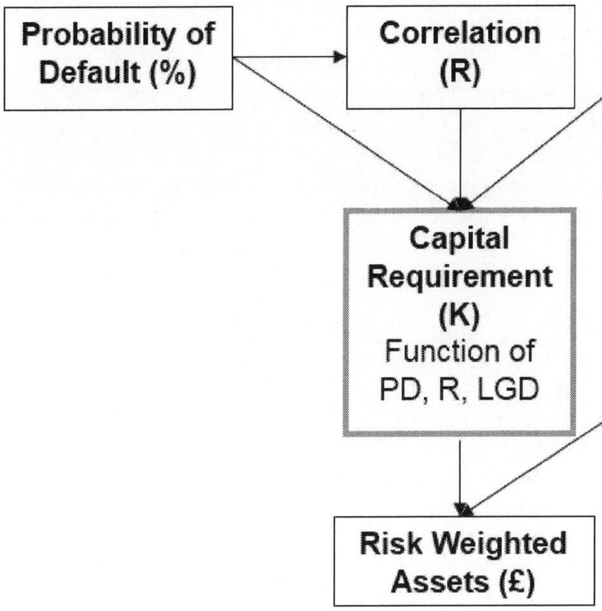

The Capital Requirement (K) is defined as a function of PD, a correlation factor (R) and LGD

$$K = LGD \times \left(\phi \left(\sqrt{\frac{1}{1-R}} \phi^{-1}(PD) + \sqrt{\frac{R}{1-R}} \phi^{-1}(0.999) \right) - PD \right) \quad (1.4)$$

where ϕ denotes the normal cumulative distribution function and ϕ^{-1} denotes the inverse cumulative distribution function. The correlation factor (R) is determined based on the portfolio being assessed. For example, for revolving retail exposures (credit cards) not in default, the correlation factor is set to 4%. A full derivation of the capital requirement can be found in Basel Committee on Banking Supervision (2004).

In practice, how do estimations of PD, LGD and EAD impact the overall capital requirements? If we take PD as 0.03, LGD as 0.5, and EAD as \$10,000, then $K(0.03, 0.5) \times (10000) = \$34.37$. If an overestimate of 10% was made on PD, then the resulting capital required would then be $K(0.033, 0.5) \times (10000) = \$36.73$, requiring an increase of 6.9% in capital (\$2.36). However if an overestimate of 10% was made on LGD, then the resulting capital required would be $K(0.03, 0.55) \times (10000) = \$37.80$, requiring an increase of 10% in capital (\$3.43).

Because LGD and EAD enter the Risk Weight Function in a linear way, it is of crucial importance to have models that estimate LGD and EAD as accurately as possible, as LGD and EAD errors are more expensive than PD errors.

1.4 SAS Software Utilized

Throughout this book, examples and screenshots aid in the understanding and practical implementation of model development. The key tools used to achieve this are Base SAS programming with SAS/STAT procedures, as well as the point-and-click interfaces of SAS Enterprise Guide and SAS Enterprise Miner. For model report generation and performance monitoring, examples are drawn from SAS Model Manager. Base SAS is a comprehensive programming language used throughout multiple industries to manage and model data. SAS Enterprise Guide (Figure 1.5) is a powerful Microsoft Windows client application that provides a guided mechanism to exploit the power of SAS and publish dynamic results throughout the organization through a point-and-click interface. SAS Enterprise Miner (Figure 1.6) is a powerful data mining tool for applying advanced modeling techniques to large volumes of data in order to achieve a greater understanding of the underlying data. SAS Model Manager (Figure 1.7) is a tool which encompasses the steps of creating, managing, deploying, monitoring, and operationalizing analytic models, ensuring the best model at the right time is in production.

Typically analysts utilize a variety of tools in their development and refinement of model building and data visualization. Through a step-by-step approach, we can identify which tool from the SAS toolbox is best suited for each task a modeler will encounter.

Figure 1.5: Enterprise Guide Interface

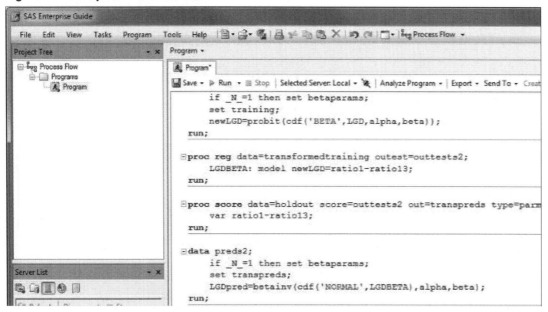

Figure 1.6: Enterprise Miner Interface

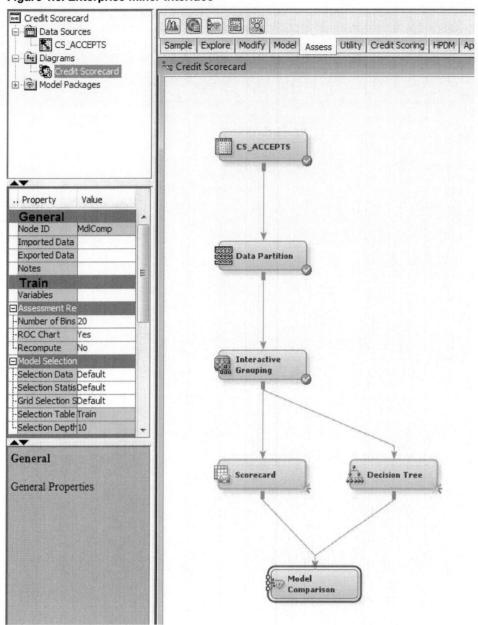

Figure 1.7: Model Manager Interface

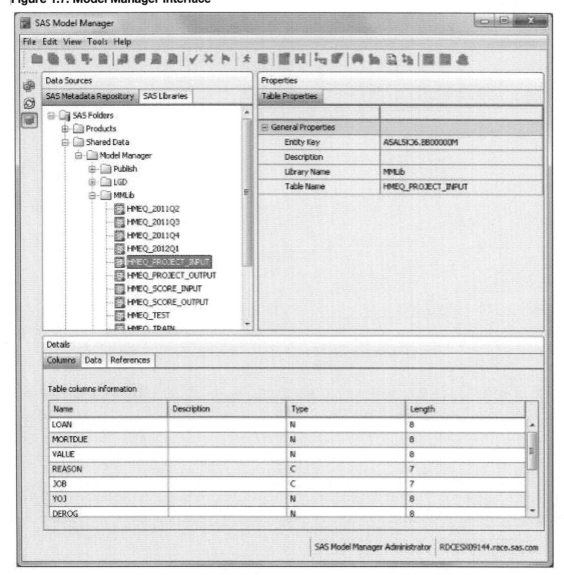

1.5 Chapter Summary

This introductory chapter explores the key concepts that comprise credit risk modeling, and how this impacts financial institutions in the form of the regulatory environment. We have also looked at how regulations have evolved over time to better account for global risks and to fundamentally prevent financial institutions from over exposing themselves to difficult market factors. To summarize, Basel defines how financial institutions calculate:

- **Expected Loss (EL)** - the means loss over 12 months
- **Unexpected Loss (UL)** - the difference between the Expected Loss and a 1 in 1000 chance level of loss
- **Risk-Weighted Assets (RWA)** - the assets of the financial institution (money lent out to customers & businesses) accounted for by their riskiness
- How much **Capital** financial institutions hold to cover these losses

Three key parameters underpin the calculation of expected loss and risk weighted assets:

- **Probability of Default (PD)** - the likelihood that a loan will not be repaid and will therefore fall into default in the next 12 months
- **Loss Given Default (LGD)** - the estimated economic loss, expressed as a percentage of exposure, which will be incurred if an obligor goes into default - in other words, LGD equals: 1 minus the recovery rate
- **Exposure At Default (EAD)** - a measure of the monetary exposure should an obligor go into default

The purpose of these regulatory requirements is to strengthen the stability of the banking system by ensuring adequate provisions for loss are made.

We have also outlined the SAS technology which will be used through a step-by-step approach to apply the theoretical information given into practical examples.

In order for financial institutions to estimate these three key parameters that underpin the calculation of EL and RWA, they must begin by utilizing the correct data. Chapter 2 covers the area of sampling and data pre-processing. In this chapter, issues such as variable selection, missing values, and outlier detection are defined and contextualized within the area of credit risk modeling. Practical applications of how these issues can be solved are also given.

1.6 References and Further Reading

Basel Committee on Banking Supervision. 2001a. *The New Basel Capital Accord.* Jan. Available at: http://www.bis.org/publ/bcbsca03.pdf.

Basel Committee on Banking Supervision. 2004. *International Convergence of Capital Measurement and Capital Standards: a Revised Framework.* Bank for International Settlements.

SAS Institute. 2002. "Comply and Exceed: Credit Risk Management for Basel II and Beyond." A SAS White Paper.

Schuermann T. 2004. "What do we know about loss given default?" Working Paper No. 04-01, Wharton Financial Institutions Center, Feb.

Chapter 2 Sampling and Data Pre-Processing

2.1 Introduction

Data is the key to unlock the creation of robust and accurate models that will provide financial institutions with valuable insight to fully understand the risks they face. However, data is often inadequate on its own and needs to be cleaned, polished, and molded into a much richer form. In order to achieve this, sampling and data pre-processing techniques can be applied in order to give the most accurate and informative insight possible.

There is an often used expression that 80% of a modeler's effort is spent in the data preparation phase, leaving only 20% for the model development. We would tend to agree with this statement; however, developing a clear, concise and logical data pre-processing strategy at the start of a project can drastically reduce this time for subsequent projects. Once an analyst knows when and where techniques should be used and the pitfalls to be aware of, their time can be spent on the development of better models that will be beneficial to the business. This chapter aims to provide analysts with this knowledge to become more effective and efficient in the data pre-processing phase by answering questions such as:

- Why is sampling and data pre-processing so important?
- What types of pre-processing are required for credit risk modeling?
- How are these techniques applied in practice?

The fundamental motivations behind the need for data cleansing and pre-processing are that data is not always in a clean fit state for practice. Often data is dirty or "noisy;" for example, a customer's age might be incorrectly recorded as 200 or their gender encoded as missing. This could purely be the result of the data collection process where human imputation error can prevail, but it is important to understand these inaccuracies in order to accurately understand and profile customers. Others examples include:

- Inconsistent data where proxy missing values are used; -999 is used to determine a missing value in one data feed, whereas 8888 is used in another data feed.
- Duplication of data; this often occurs where disparate data sources are collated and merged, giving an unclear picture of the current state of the data.

- Missing values and extreme outliers; these can be treated, removed or used as they are in the modeling process. For example, some techniques, such as decision trees (Figure 2.1), can cope with missing values and extreme outliers more effectively than others. Logistic regression cannot handle missing values without excluding observations or applying imputation. (This concept will be discussed in more detail later in the chapter).

Figure 2.1: Handling Missing Values in a Decision Tree

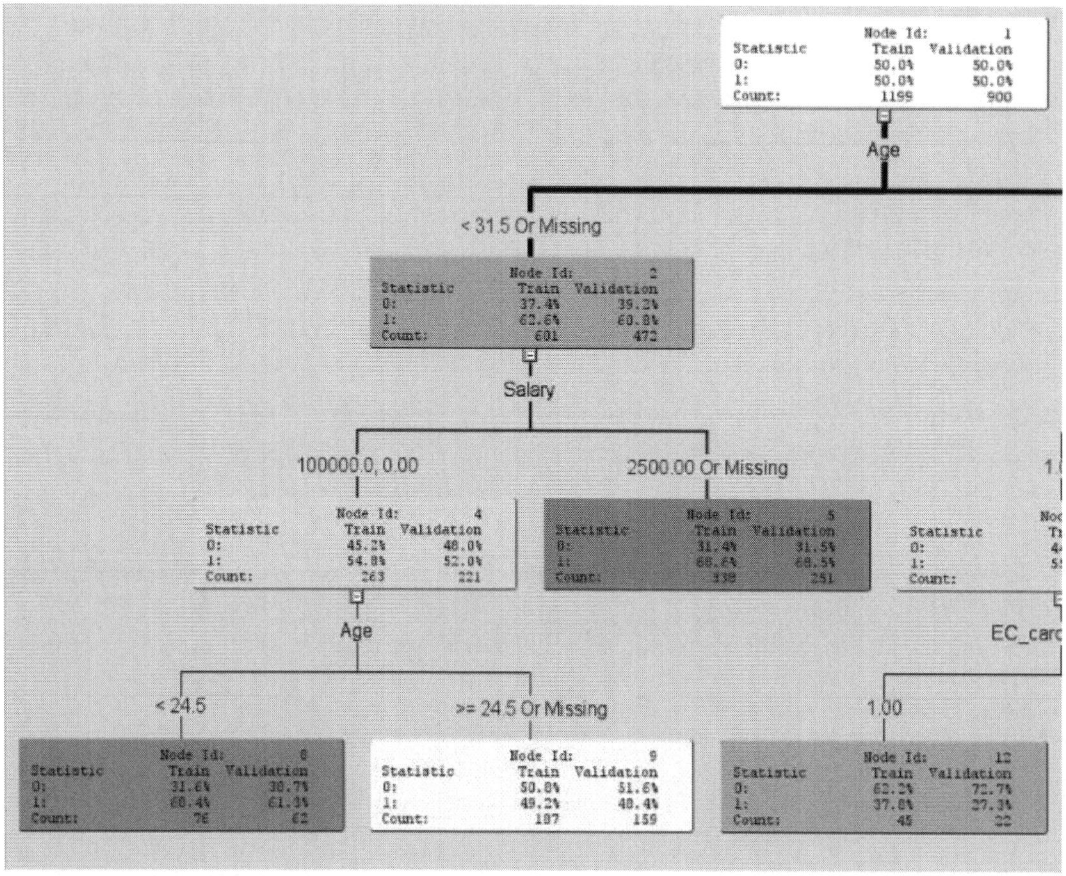

A well-worn term in the field of data modeling is "garbage in, garbage out," meaning that if the data you have coming into your model is incorrect, inconsistent, and dirty then an inaccurate model will result no matter how much time is spent on the modeling phase. It is also worth noting that this is by no means an easy process; as mentioned above, data pre-processing tends to be time consuming. The rule of thumb is to spend 80% of your time preparing the data to 20% of the time actually spent on building accurate models.

Data values can also come in a variety of forms. The types of variables typically utilized within a credit risk model build fall into two distinct categories; Interval and Discrete.

Interval variables (also termed continuous) are variables that typically can take any numeric value from $-\infty$ to ∞. Examples of interval variables are any monetary amount such as current balance, income, or amount outstanding. Discrete variables can be both numeric and non-numeric but contain distinct separate values that are not continuous. Discrete variables can be further split into three categories: Nominal, Ordinal, and Binary. Nominal variables contain no order between the values, such as marital status (Married, Single, Divorced, etc.) or Gender (Male, Female, and Unknown). Ordinal variables share the same properties as Nominal variables; however, there is a ranked ordering or hierarchy between the variables for example rating grades (AAA, AA, A…). Binary variables contain two distinct categories of data, for example, if a customer has defaulted (bad category) or not defaulted (good category) on a loan.

When preparing data for use with SAS Enterprise Miner, one must first identify how the data will be treated. Figure 2.2 shows how the data is divided into categories.

Figure 2.2: Enterprise Miner Data Source Wizard

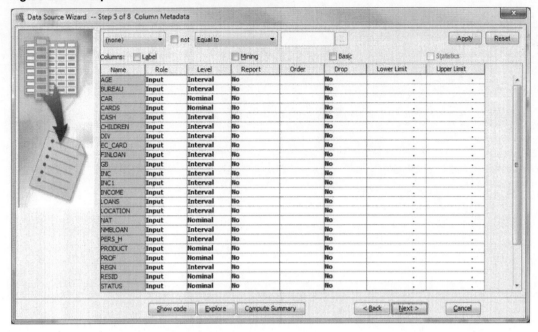

The Enterprise Miner Data Source Wizard automatically assigns estimated levels to the data being brought into the workspace. This should then be explored to determine whether the correct levels have been assigned to the variables of interest. Figure 2.3 shows how you can explore the variable distributions.

Figure 2.3: Variable Distributions Displayed in the Explore Tab of the Enterprise Miner Data Source Wizard

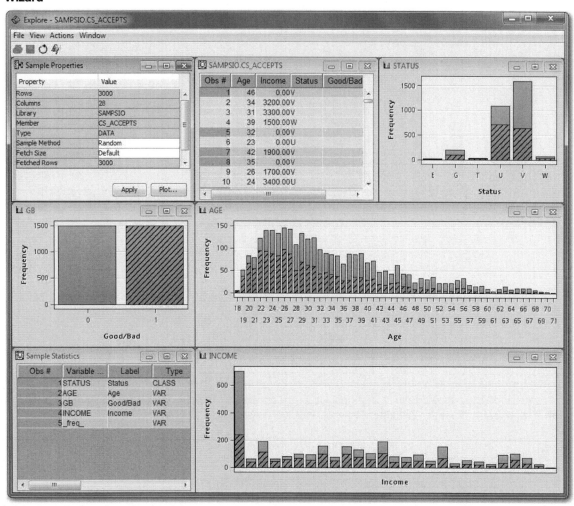

This chapter will highlight the key data pre-processing techniques required in a credit risk modeling context before a modeling project is undertaken. The areas we will focus on are:

- Sampling
- Variable selection and correlation analysis
- Missing values
- Outlier detection and treatment
- Clustering & Segmentation

Throughout this chapter, we will utilize SAS software capabilities and show with examples how each technique can be achieved in practice.

2.2 Sampling and Variable Selection

In this section, we discuss the topics of data sampling and variable selection in the context of credit risk model development. We explore how sampling methodologies are chosen as well as how data is partitioned into separate roles for use in model building and validation. The techniques that are available for variable selection are given as well as a process for variable reduction in a model development project.

2.2.1 Sampling

Sampling is the process of extracting a relevant number of historical data cases (in this case, credit card applicants) from a larger database. From a credit risk perspective, the extracted sample needs to be fit for the type of business analysis being undertaken. It is often impractical to build a model on the full population, as this can be time-consuming and involve a high volume of processing. It is therefore important to first determine what population is required for the business problem being solved. Of equal importance is the timescales for which the sample is taken; for example, what window of the data do you need to extract? Do you want to focus on more recent data or have a longer sample history? Another consideration to make is the distribution of the target you wish to estimate. With regards to estimating whether a customer will default or not, some portfolios will exhibit a large class imbalance with a 1% default rate to 99% non-default rate. When class imbalances are present, techniques such as under-sampling the non-defaults or over-sampling the defaulting observations may be needed to address this imbalance.

For successful data sampling, samples must be from a normal business period to give as accurate a picture as possible of the target population that is being estimated. For example, considerations around global economic conditions and downturns in the economy must be taken into account when identifying a normal business period. A performance window must also be taken that is long enough to stabilize the bad rate over time (12 - 18 months). Examples of sampling problems experienced in credit risk modeling include:

- In application scoring – information regarding historic good/bad candidates is only based upon those candidates who were offered a loan (known good/bads) and does not take into account those candidates that were not offered a loan in the first instance. Reject Inference adjusts for this by inferring how those candidates not offered a loan would have behaved based on those candidates that we know were good or bad. An augmented sample is then created with the known good/bads plus the inferred good/bads.

- In behavioral scoring – seasonality can play a key role depending upon the choice of the observation point. For example, credit card utilization rates tend to increase around seasonal holiday periods such as Christmas.

In SAS, either the use of **proc surveyselect** in SAS/STAT or the **Sample node** in SAS Enterprise Miner can be utilized to take stratified or simple random samples of larger volumes of data. Once the overall sample of data required for analysis has been identified, further data sampling (data partitioning) should also be undertaken so as to have separate data sources for model building and model validation. In SAS Enterprise Miner, data can be further split into three separate data sources: Training, Validation and Test sets. The purpose of this is to make sure models are not being over-fit on a single source, but can generalize to unseen data and thus to real world problems. A widely used approach is to split the data into two-thirds training and one-third validation when only two sets are required; however, actual practices vary depending on internal model validation team recommendations, history, and personal preferences. Typically, splits such as 50/50, 80/20, and 70/30 are used. A test set may be used for further model tuning, such as with neural network models. Figure 2.4 shows an example of an Enterprise Miner **Data Partition node** and property panel with a 70% randomly sampled Training set and 30% Validation set selected.

Figure 2.4: Enterprise Miner Data Partition Node and Property Panel (Sample Tab)

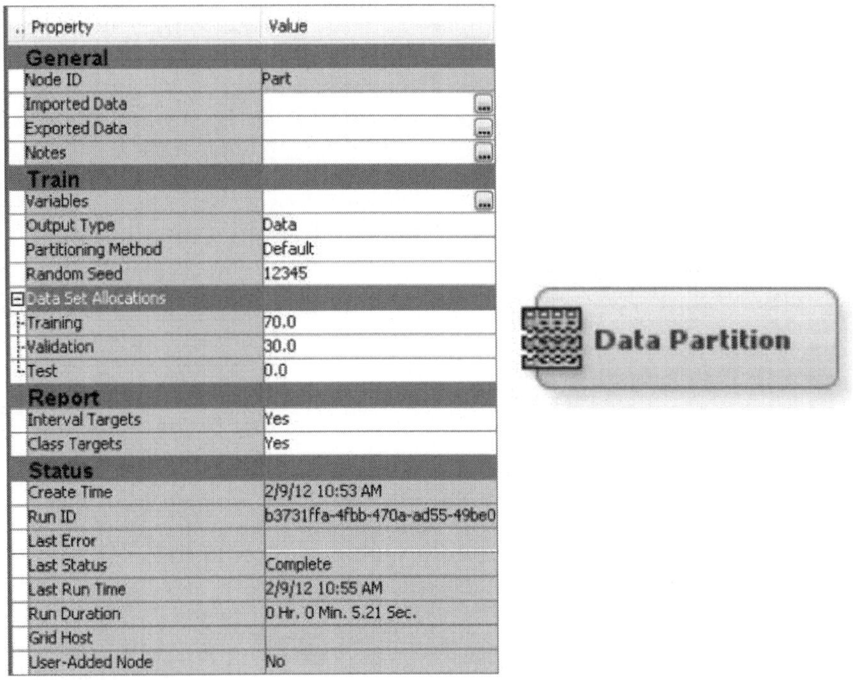

Property	Value
General	
Node ID	Part
Imported Data	
Exported Data	
Notes	
Train	
Variables	
Output Type	Data
Partitioning Method	Default
Random Seed	12345
Data Set Allocations	
Training	70.0
Validation	30.0
Test	0.0
Report	
Interval Targets	Yes
Class Targets	Yes
Status	
Create Time	2/9/12 10:53 AM
Run ID	b3731ffa-4fbb-470a-ad55-49be0
Last Error	
Last Status	Complete
Last Run Time	2/9/12 10:55 AM
Run Duration	0 Hr. 0 Min. 5.21 Sec.
Grid Host	
User-Added Node	No

2.2.2 Variable Selection

Organizations often have access to a large number of potential variables that could be used in the modeling process for a number of different business questions. Herein lies a problem: out of these potentially thousands of variables, which ones are useful to solve a particular issue? Variable selection under its many names (input selection, attribute selection, feature selection) is a process in which statistical techniques are applied at a variable level to identify those that have the most descriptive power for a particular target.

There are a wide number of variable selection techniques available to practitioners, including but not exclusive to:

- Correlation analysis (Pearson's p, Spearman's r and Kendall's tau)
- Stepwise regression
- Information value based selection
- Chi-squared analysis
- Variable clustering
- Gain/entropy
- ANOVA analysis etc.

A good variable selection subset should only contain variables predictive of a target variable yet un-predictive of each other (Hall and Smith, 1998). By conducting variable selection, improvements in model performance and processing time can be made.

In terms of developing an input selection process, it is common practice to first use a quick filtering process to reduce the overall number of variables to a manageable size. The use of the **Variable Selection node** or **Variable Clustering node** in the Explore tab of SAS Enterprise Miner allows the quick reduction of variables independent of the classification algorithm (linear regression) used in the model development phase. The **Variable Clustering node** in particular is an extremely powerful tool for identifying any strong correlations or covariance that exists within the input space. This node will identify associated groups within the input space and either select a linear combination of the variables in each cluster or the best variable in each cluster that have the minimum R-square ratio values. In the context of credit risk modeling, the most appropriate strategy is to select the best variable from a clustered group in order to retain a clearer relational understanding of the inputs to the dependent variable. In most cases, you will also want to force additional variables through even if they are not selected as best, and this can be achieved using the **Interaction Selection** option on the **Variable Clustering node**.

Once the first reduction of variables has been made, forward/backward/stepwise regression can be used to further determine the most predictive variables based on their p-values. This two stage approach allows for a dual variable reduction process which also speeds up model processing times, as a preliminary reduction has been made before the model is built.

Table 2.1 details the typical variable selection techniques that can be used relative to the type of input and target.

Table 2.1: Variable Selection Techniques

	Continuous Target (LGD)	Discrete Target (PD)
Interval input (Salary)	Pearson correlation	Fisher score
Class input (Gender)	Fisher score ANOVA analysis	Chi-squared analysis Cramer's V Information Value Gain/Entropy

A number of these variable selection techniques are utilized in the forthcoming chapters, with full examples and explanations given.

Additional criteria to consider when conducting variable selection include the interpretability of the input you wish to use for the model. Do inputs display the expected sign when compared to the target? For example, as exposure at default increases, we would expect a positive (+) relationship to utilization rates. If a variable cannot be clearly interpreted in terms of its relationship to the target, it may be rendered unusable.

Legal concerns should also be taken into account, as different countries have different legislation with regards to the use of variables for discriminatory purposes (such as nationality and ethnicity). It is also important to note that business judgment is widely used in the decision-making process. To make informed and appropriate decisions as to which variables will be used, practitioners must use their own business judgment alongside the statistical output the above tests offer.

2.3 Missing Values and Outlier Treatment

2.3.1 Missing Values

A perennial issue of data modeling and analysis is the presence of missing values within data. This can result from a number of causes, such as human imputational error, non-disclosure of information such as gender, and non-applicable data where a particular value is unknown for a customer. In dealing with missing values in the data, one must first decide whether to keep, delete, or replace missing values. It is important to note that even though a variable contains missing values, this information may be important in itself. For example, it could indicate fraudulent behavior if customers are purposefully withholding certain pieces of information. To mitigate for this, an additional category can be added to the data for missing values (such as missing equals -999) so that this information is stored and can be used in the modeling process. The deletion of missing values is usually only appropriate when there are too many missing values in a variable for any useful information to be gained. A choice must then be made as to whether horizontal or vertical deletion is conducted. Table 2.2 depicts a case where 70% of the Credit Score variable is missing and 60% of Account ID 01004 is missing (shaded grey). It may make sense in this context to remove both the Credit Score variable in its entirety and Account ID observation 01004.

Table 2.2: Identification of Missing Values

Account ID	Age	Marital Status	Gender	Credit Score	Good/Bad
01001	21	Single	M	550	BAD
01002	45	Single	F	?	GOOD
01003	?	Married	F	?	GOOD
01004	30	?	?	?	GOOD
01005	26	Single	M	?	BAD
01006	32	Single	F	650	GOOD
01007	60	Divorced	F	?	BAD
01008	45	Single	M	620	GOOD
01009	50	?	M	?	BAD
01010	22	Single	M	?	BAD

The replacement of missing values involves using imputation procedures to estimate what the missing value is likely to be. It is important, however, to be consistent in this process throughout the model development phase and model usage so as not to bias the model.

It is worth noting that missing values can play a different role in the model build phase to the application for a loan phase. If historic data observations have missing values, it would usually warrant the imputation of values and fitting models with and without the imputed values, and hopefully, the final model will be more robust in the presence of missing values. But if a loan applicant does not provide required information in the application phase, you can request the values again, and if they cannot provide it, this might be sufficient cause to reject the loan application.

The **Impute node** (shown in Figure 2.5) can be utilized in SAS Enterprise Miner for this, which includes the following imputation methods for missing class and interval variable values (Table 2.3):

Figure 2.5: Enterprise Miner Imputation Node (Modify Tab)

Table 2.3: Imputation Techniques

Discrete (Categorical) Variables	Interval (Continuous) Variables
Input/Target	*Input/Target*
Count	Mean
Default Constant Value	Median
Distribution	Midrange
Tree (only for inputs)	Distribution
Tree Surrogate (only for inputs)	Tree (only for inputs)
	Tree Surrogate (only for inputs)
	Mid-Minimum Spacing
	Tukey's Biweight
	Huber
	Andrew's Wave
	Default Constant

(A detailed explanation of the imputation techniques for class and interval variables detailed in Table 2.3 can be found in the SAS Enterprise Miner help file).

From a credit risk modeling perspective, it is important to understand the implications of assuming the value of a missing attribute. Care must be taken in the application of these techniques with the end goal to enrich the data and improve the overall model. It is often a worthwhile activity to create both an imputed model and a model without imputation to understand the differences in performance. Remember, modeling techniques such as decision trees inherently account for missing values, and missing value treatment is conducted as part of the tree growth. There is also argument to say binning strategies should be implemented prior to model building to deal with missing values within continuous variables (Van Berkel & Siddiqi, 2012). The topic of binning prior to modeling PD and LGD values will be discussed in more detail in their respective chapters.

Note: Although more complex techniques are available for missing value imputation, in practice, this does not usually result in any substantial improvements in risk model development.

2.3.2 Outlier Detection

The process of outlier detection aims to highlight extreme or unusual observations, which could have resulted from data entry issues or more generally noise within the data. Outliers take the form of both valid observations (in the analysis of incomes, a CEO's salary may stand out as an outlier when analyzing general staff's pay) and invalid observations, for example, a negative integer being recorded for a customer's age. In this regard, precaution must be taken in the handling of identified outliers, as outlier detection may require treatment or removal from the data completely. For example in a given range of loan sizes, the group with the highest annual income (or the most years of service in their current job) might be the best loan applicants and have the lowest probability of default.

In order to understand and smooth outliers, a process of outlier detection should be undertaken. SAS Enterprise Miner incorporates several techniques to achieve this. Through the use of the **Filter node** (Figure 2.6) a variety of automated and manual outlier treatment techniques can be applied.

Figure 2.6: Enterprise Miner Filter node (Sample Tab)

For Interval Variables, the filtering methods available are:

- Mean Absolute Deviation (MAD)
- User-Specified Limits
- Metadata Limits
- Extreme Percentiles
- Modal Centre
- Standard Deviations from the Mean

For Discrete Variables, the filtering methods available are:

- Rare Values (Count)
- Rare Values (Percentage)

For best results, data should be visualized and outliers should first be understood before the application of outlier techniques. The user should decide which model is to be used and how outliers can affect the interpretability of the model output. A full documentation of each of these techniques can be found in the Help section of Enterprise Miner.

2.4 Data Segmentation

In the context of credit risk modeling, it is important to understand the different tranches of risk within a portfolio. A key mechanism to better understand the riskiness of borrowers is the process of segmentation (clustering), which categorizes them into discrete buckets based on similar attributes. For example, by separating out those borrowers with high credit card utilization rates from those with lower utilization rates, different approaches can be adopted in terms of targeted marketing campaigns and risk ratings.

When looking to develop a credit risk scorecard, it is often inappropriate to apply a single scorecard to the whole population. Different scorecards are required to treat each disparate segment of the overall portfolio separately. There are three key reasons why financial institutions would want to do this (Thomas, Ho, and Scherer, 2001):

- Operational
- Strategic
- Variable Interactions

In terms of an operation perspective on segmentation, new customers wishing to take out a financial product from a bank could be given a separate scorecard, because the characteristics in a standard scorecard do not make operational sense for them. From a strategic perspective, banks may wish to adopt particular strategies for defined segments of customers; for example, if a financial institution's strategy is to increase the number of student customers, a lower cutoff score might be defined for this segment. Segmentation is also not purely based upon observations. If a certain variable interacts strongly with a number of other highly correlated variables, it can often be more appropriate to segment the data on a particular variable instead.

In order to separate data through segmentation, two main approaches can be applied:

- Experience or business rules based
- Statistically based

The end goal in any segmentation exercise is to garner a better understanding of your current population and to create a more accurate or more powerful prediction of their individual attributes. From an experience or business rules based approach, experts within the field can advise where the best partitioning of customers lies based on their business knowledge. The problem that can arise from this, however, is that this decision making is not as sound as a decision based on the underlying data, and can lead to inaccurate segments being defined. From a statistically based approach, clustering algorithms such as decision trees and k-means clustering can be applied to data to identify the best splits. This enhances the decision making process by coupling a practitioner's experience of the business with the available data to better understand the best partitioning within the data.

In SAS Enterprise Miner, there are a variety of techniques which can be used in the creation of segmented models. We provide an overview of decision trees, where both business rules and statistics can be combined, and then go on to look at k-means based segmentations.

2.4.1 Decision Trees for Segmentation

Decision trees (Figure 2.7) represent a modular supervised segmentation of a defined data source created by applying a series of rules which can be determined both empirically and by business users. Supervised segmentation approaches are applicable when a particular target or flag is known. Each rule assigns an observation to a segment based on the value of one input. One rule is applied after another, resulting in a hierarchy (*tree*) of segments within segments (called *nodes*). The original segment contains the entire data set and is called the *root node* of the tree; in the example below, this is our total population of *known good bads* (KGBs). A node with all its successors forms a branch of the node that created it. The final nodes are called *leaves* (Figure 2.8). Decision trees have both predictive modeling applications as well as segmentation capabilities, and the predictive aspect will be discussed in detail in Chapter 4.

Figure 2.7: Enterprise Miner Decision Tree Node (Model Tab)

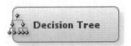

Figure 2.8: A Conceptual Tree Design for Segmentation

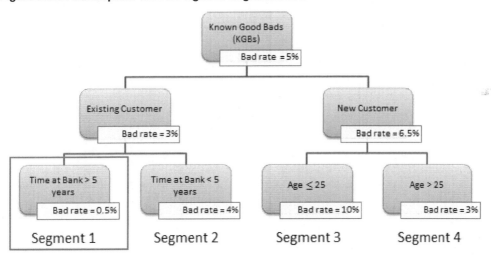

Figure 2.8 illustrates that through the process of decision tree splits, we can determine a tranche of existing customers who have been with the bank longer than five years are those with the smallest bad rate. Through the identification of visual segments, it becomes much easier to disseminate this information throughout the organization. Clearly defined segments allow for the correct identification of risk profiles and can lead to business decisions in terms of where to concentrate collections and recover strategies. Segment 1, shown in Figure 2.8, would give the following rule logic which could be used to score new and existing customers:

```
*----------------------------------------------------------*

  Segment = 1

*----------------------------------------------------------*

if EXISTING_CUSTOMER = Y
```

AND TAB > 5

then

Tree Node Identifier = 1

Predicted: Bad_Rate = 0.005

The visual nature of decision trees also makes it easier for practitioners to explain interactions and conclusions that can be made from the data to decision makers, regulators, and model validators. Because simple explainable models are preferred over more complex models, this is essential from a regulatory perspective.

Note: The use of decision trees for modeling purposes will be discussed further in Chapter 3 and Chapter 4.

As an example, a challenger financial institution in the UK wanted to differentiate their credit line offerings based on the development of bespoke segmented scorecards. Through the application of a data-driven unsupervised learning segmentation model in SAS, we were able to achieve a 3.5% portfolio decrease in bad rate due to a better differentiation of their customer base.

2.4.2 K-Means Clustering

The SAS Enterprise Miner **Cluster node** (Figure 2.9) can be used to perform observation clustering, which can create segments of the full population. K-means clustering is a form of unsupervised learning, which means a target or flag is not required and the learning process attempts to find appropriate groupings based in the interactions within the data. Clustering places objects into groups or clusters suggested by the data. The objects in each cluster tend to be similar to each other in some sense, and objects in different clusters tend to be dissimilar. If obvious clusters or groupings could be developed prior to the analysis, then the clustering analysis could be performed by simply sorting the data.

Figure 2.9: Enterprise Miner Cluster Node (Explore Tab)

The Cluster node performs a disjoint cluster analysis on the basis of distances computed from one or more quantitative variables. The observations are divided into clusters such that every observation belongs to one and only one cluster; the clusters do not form a tree structure as they do in the **Decision Tree node**. By default, the Cluster node uses Euclidean distances, so the cluster centers are based on least squares estimation. This kind of clustering method is often called a k-means model, since the cluster centers are the means of the observations assigned to each cluster when the algorithm is run to complete convergence. Each iteration reduces the least squares criterion until convergence is achieved. The objective function is to minimize the sum of squared distances of the points in the cluster to the cluster mean for all the clusters, in other words, to assign the observations to the cluster they are closest to. As the number of clusters increases, observations can then be assigned to different clusters based on their centrality.

To understand the clusters created by this node, a **Segment Profile node** (Figure 2.10) should be appended to the flow. This profile node enables analysts to examine the clusters created and examine the distinguishing factors that drive those segments. This is important to determine what makes up distinct groups of the population in order to identify particular targeting or collection campaigns. Within a credit risk modeling context, the aim is to understand the risk profiles of customer groups. For example, you might want to identify whether age or household income is driving default rates in a particular group. The blue bars in the segment profile output (Figure 2.11) display the distributions for an actual segment, whereas the red bars display the distribution for the entire population.

Figure 2.10: Enterprise Miner Segment Profile Node (Assess Tab)

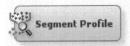

Figure 2.11: Segment Profile Output

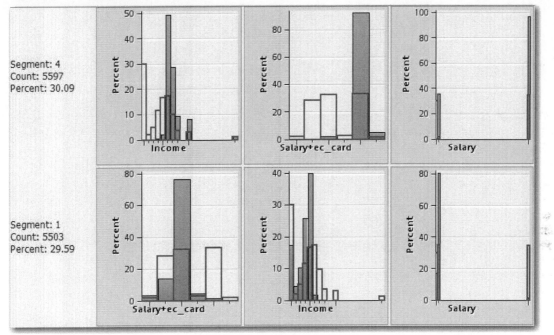

2.5 Chapter Summary

In this chapter, we have covered the key sampling and data pre-processing techniques one should consider in the data preparation stage prior to model building. From a credit risk modeling perspective, it is crucial to have the correct view of the data being used for modeling. Through the application of data pre-processing techniques such as missing value imputation, outlier detection, and segmentation, analysts can achieve more robust and precise risk models. We have also covered the issue of dimensionality; where a number of variables are present, variable selection processes can be applied in order to speed up processing time and reduce variable redundancy.

Building upon the topics we have covered so far, the following chapter, Chapter 3, details the theory and practical aspects behind the creation of a probability of default model. This focuses on standard and novel modeling techniques and shows how each of these can be used in the estimation of PD. The development of an application and behavioral scorecard using SAS Enterprise Miner demonstrates a practical implementation.

2.6 References and Further Reading

Hall, M.A. and Smith L.A. 1998. "Practical Feature Subset Selection for Machine Learning." Proc Australian Computer Science Conference, 181-191. Perth, Australia.

Thomas, L.C., Ho, J. and Scherer, W.T. 2001. "Time Will Tell: Behavioral Scoring and the Dynamics of Consumer Credit Assessment." IMA Journal of Management Mathematics, 12, (1), 89-103. (doi:10.1093/imaman/12.1.89).

Van Berkel, A. and Siddiqi, N. 2012. "Building Loss Given Default Scorecard Using Weight of Evidence Bins in SAS Enterprise Miner." Proceedings of the SAS Global Forum 2012 Conference, Paper 141-2012. Cary, NC: SAS Institute Inc.

Chapter 3 Development of a Probability of Default (PD) Model

3.1 Overview of Probability of Default

Over the last few decades, the main focus of credit risk modeling has been on the estimation of the Probability of Default (PD) on individual loans or pools of transactions. PD can be defined as the likelihood that a loan will not be repaid and will therefore fall into default. A default is considered to have occurred with regard to a particular obligor (a customer) when either or both of the two following events have taken place:

1. The bank considers that the obligor is unlikely to pay its credit obligations to the banking group in full (for example, if an obligor declares bankruptcy), without recourse by the bank to actions such as

realizing security (if held) (for example, taking ownership of the obligor's house, if they were to default on a mortgage).

2. The obligor has missed payments and is past due for more than 90 days on any material credit obligation to the banking group (Basel, 2004).

In this chapter, we look at how PD models can be constructed both at the point of application, where a new customer applies for a loan, and at a behavioral level, where we have information with regards to current customers' behavioral attributes within the cycle. A distinction can also be made between those models developed for retail credit and corporate credit facilities in the estimation of PD. As such, this overview section has been sub-divided into three categories distinguishing the literature for retail credit (Section 3.1.1), corporate credit (Section 3.1.2) and calibration (Section 3.1.3).

Following this section we focus on retail portfolios by giving a step-by-step process for the estimation of PD through the use of SAS Enterprise Miner and SAS/STAT. At each stage, examples will be given using real world financial data. This chapter will also develop both an application and behavioral scorecard to demonstrate how PD can be estimated and related to business practices. This chapter aims to show how parameter estimates and comparative statistics can be calculated in Enterprise Miner to determine the best overall model. A full description of the data used within this chapter can be found in the appendix section of this book.

3.1.1 PD Models for Retail Credit

Credit scoring analysis is the most well-known and widely used methodology to measure default risk in consumer lending. Traditionally, most credit scoring models are based on the use of historical loan and borrower data to identify which characteristics can distinguish between defaulted and non-defaulted loans (Giambona & Iancono, 2008). In terms of the credit scoring models used in practice, the following list highlights the five main traditional forms:

1. Linear probability models (Altman, 1968);
2. Logit models (Martin, 1977);
3. Probit models (Ohlson, 1980);
4. Multiple discriminant analysis models;
5. Decision trees.

The main benefits of credit scoring models are their relative ease of implementation and their transparency, as opposed to some more recently proposed "black-box" techniques such as Neural Networks and Least Square Support Vector Machines. However there is merit in a comparison approach of more non-linear black-box techniques against traditional techniques to understand the best potential model that can be built.

Since the advent of the Basel II capital accord (Basel Committee on Banking Supervision, 2004), a renewed interest has been seen in credit risk modeling. With the allowance under the internal ratings-based approach of the capital accord for organizations to create their own internal ratings models, the use of appropriate modeling techniques is ever more prevalent. Banks must now weigh the issue of holding enough capital to limit insolvency risks against holding excessive capital due to its cost and limits to efficiency (Bonfim, 2009).

3.1.2 PD Models for Corporate Credit

With regards to corporate PD models, West (2000) provides a comprehensive study of the credit scoring accuracy of five neural network models on two corporate credit data sets. The neural network models are then benchmarked against traditional techniques such as linear discriminant analysis, logistic regression, and k-nearest neighbors. The findings demonstrate that although the neural network models perform well, more simplistic, logistic regression is a good alternative with the benefit of being much more readable and understandable. A limiting factor of this study is that it only focuses on the application of additional neural network techniques on two relatively small data sets, and doesn't take into account larger data sets or other machine learning approaches. Other recent work worth reading on the topic of PD estimation for corporate credit can be found in Fernandes (2005), Carling et al (2007), Tarashev (2008), Miyake and Inoue (2009), and Kiefer (2010).

3.1.3 PD Calibration

The purpose of PD calibration is to assign a default probability to each possible score or rating grade values. The important information required for calibrating PD models includes:

- The PD forecasts over a rating class and the credit portfolio for a specific forecasting period.
- The number of obligors assigned to the respective rating class by the model.
- The default status of the debtors at the end of the forecasting period.

(Guettler and Liedtke, 2007)

It has been found that realized default rates are actually subject to relatively large fluctuations, making it necessary to develop indicators to show how well a rating model estimates the PDs (Guettler and Liedtke, 2007). Tasche recommends that traffic light indicators could be used to show whether there is any significance in the deviations of the realized and forecasted default rates (2003). The three traffic light indicators (green, yellow, and red) identify the following potential issues:

- A green traffic light indicates that the true default rate is equal to, or lower than, the upper bound default rate at a low confidence level.
- A yellow traffic light indicates the true default rate is higher than the upper bound default rate at a low confidence level and equal to, or lower than, the upper bound default rate at a high confidence level.
- Finally a red traffic light indicates the true default rate is higher than the upper bound default rate at a high confidence level. (Tasche, 2003 via Guettler and Liedtke, 2007)

3.2 Classification Techniques for PD

Classification is defined as the process of assigning a given piece of input data into one of a given number of categories. In terms of PD modeling, classification techniques are applied to estimate the likelihood that a loan will not be repaid and will fall into default. This requires the classification of loan applicants into two classes: good payers (those who are likely to keep up with their repayments) and bad payers (those who are likely to default on their loans).

In this section, we will highlight a wide range of classification techniques that can be used in a PD estimation context. All of the techniques can be computed within the SAS Enterprise Miner environment to enable analysts to compare their performance and better understand any relationships that exist within the data. Further on in the chapter, we will benchmark a selection of these to better understand their performance in predicting PD. An empirical explanation of each of the classification techniques applied in this chapter is presented below. This section will also detail the basic concepts and functioning of a selection of well-used classification methods.

The following mathematical notations are used to define the techniques used in this book. A scalar x is denoted in normal script. A vector X is represented in boldface and is assumed to be a column vector. The corresponding row vector X^T is obtained using the transpose T. Bold capital notation is used for a matrix $\mathbf{X}$. The number of independent variables is given by n and the number of observations (each corresponding to a credit card default) is given by l. The observation i is denoted as $\mathbf{x}_i$ whereas variable j is indicated as x_j.

The dependent variable y (the value of PD, LGD or EAD) for observation i is represented as y_i. P is used to denote a probability.

3.2.1 Logistic Regression

In the estimation of PD, we focus on the binary response of whether a creditor turns out to be a good or bad payer (non-defaulter vs. defaulter). For this binary response model, the response variable y can take on one of two possible values: $y = 1$ if the customer is a bad payer; $y = 0$ if they are a good payer. Let us assume that

$\mathbf{X}$ is a column vector of M explanatory variables and $\pi = \mathrm{P}(y = 1 \mid \mathbf{x})$ is the response probability to be modeled. The logistic regression model then takes the form:

$$\mathrm{logit}(\pi) \equiv \log\left(\frac{\pi}{1-\pi}\right) = \alpha + \beta^T \mathbf{x} \quad (3.1)$$

where α is the intercept parameter and β^T contains the variable coefficients (Hosmer and Stanley, 2000).

The cumulative logit model (Walker and Duncan, 1967) is simply an extension of the binary two-class logit model which allows for an ordered discrete outcome with more than 2 levels $(k > 2)$:

$$P(\mathrm{class} \le j) = \frac{1}{1 + e^{-\left(d_j + b_1 x_1 + b_2 x_2 + \ldots + b_n x_n\right)}} \quad (3.2)$$

$$j = 1, 2, \ldots, k-1$$

The cumulative probability, denoted by $P(\mathrm{class} \le j)$, refers to the sum of the probabilities for the occurrence of response levels up to and including the j th level of y. The main advantage of logistic regression is the fact that it is a non-parametric classification technique, as no prior assumptions are made with regard to the probability distribution of the given attributes.

This approach can be formulated within SAS Enterprise Miner using the **Regression node** (Figure 3.1) within the Model tab.

Figure 3.1: Regression Node

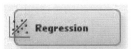

The **Regression node** can accommodate for both linear (interval target) and logistic regression (binary target) model types.

3.2.2 Linear and Quadratic Discriminant Analysis

Discriminant analysis assigns an observation to the response y ($y \in \{0,1\}$) with the largest posterior probability; in other words, classify into class 0 if $p(0\,|\,\mathbf{x}) > p(1\,|\,\mathbf{x})$, or class 1 if the reverse is true. According to Bayes' theorem, these posterior probabilities are given by:

$$p(y\,|\,\mathbf{x}) = \frac{p(\mathbf{x}\,|\,y)p(y)}{p(\mathbf{x})} \quad (3.3)$$

Assuming now that the class-conditional distributions $p(\mathbf{x}\,|\,y=0)$, $p(\mathbf{x}\,|\,y=1)$ are multivariate normal distributions with mean vector $\boldsymbol{\mu}_0$, $\boldsymbol{\mu}_1$, and covariance matrix $\boldsymbol{\Sigma}_0$, $\boldsymbol{\Sigma}_1$, respectively, the classification rule becomes classify as $y=0$ if the following is satisfied:

$$\left(\mathbf{x}-\boldsymbol{\mu}_0\right)^T \sum_0^{-1}\left(\mathbf{x}-\boldsymbol{\mu}_0\right) \; - \; \left(\mathbf{x}-\boldsymbol{\mu}_1\right)^T \sum_1^{-1}\left(\mathbf{x}-\boldsymbol{\mu}_1\right)$$
$$< 2\left(\log\left(P\left(y=0\right)-\log\left(P\left(y=1\right)\right)\right)\right)+\log\left|\boldsymbol{\Sigma}_1\right|-\log\left|\boldsymbol{\Sigma}_0\right| \quad (3.4)$$

Linear discriminant analysis (LDA) is then obtained if the simplifying assumption is made that both covariance matrices are equal ($\boldsymbol{\Sigma}_0 = \boldsymbol{\Sigma}_1 = \boldsymbol{\Sigma}$), which has the effect of cancelling out the quadratic terms in the expression above.

SAS Enterprise Miner does not contain an LDA or QDA node as standard; however, SAS/STAT does contain the procedural logic to compute these algorithms in the form of **proc discrim**. This approach can be formulated within SAS Enterprise Miner using a SAS code node, or the underlying code can be utilized to develop an Extension Node (Figure 3.2) in SAS Enterprise Miner.

Figure 3.2: LDA Node

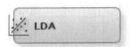

More information on creating bespoke extension nodes in SAS Enterprise Miner can be found by typing "Development Strategies for Extension Nodes" into the http://support.sas.com/ website. Program 3.1 demonstrates an example of the code syntax for developing an LDA model on the example data used within this chapter.

Program 3.1: LDA Code

```
PROC DISCRIM DATA=&EM_IMPORT_DATA WCOV PCOV CROSSLIST
        WCORR PCORR Manova testdata=&EM_IMPORT_VALIDATE testlist;
        CLASS %EM_TARGET;
        VAR %EM_INTERVAL;
run;
```

This code could be used within a **SAS Code node** after a Data Partition node using the Train set (&EM_IMPORT_DATA) to build the model and the Validation set (&EM_IMPORT_VALIDATE) to validate the model. The %EM_TARGET macro identifies the target variable (PD) and the %EM_INTERVAL macro identifies all of the interval variables. The class variables would need to be dummy encoded prior to insertion in the VAR statement.

Note: The **SAS Code node** enables you to incorporate new or existing SAS code into process flow diagrams that were developed using SAS Enterprise Miner. The SAS Code node extends the functionality of SAS Enterprise Miner by making other SAS System procedures available for use in your data mining analysis.

3.2.3 Neural Networks

Neural networks (NN) are mathematical representations modeled on the functionality of the human brain (Bishop, 1995). The added benefit of a NN is its flexibility in modeling virtually any non-linear association between input variables and the target variable. Although various architectures have been proposed, this section focuses on probably the most widely used type of NN, the Multilayer Perceptron (MLP). A MLP is typically composed of an input layer (consisting of neurons for all input variables), a hidden layer (consisting of any number of hidden neurons), and an output layer (in our case, one neuron). Each neuron processes its inputs and transmits its output value to the neurons in the subsequent layer. Each of these connections between neurons is assigned a weight during training. The output of hidden neuron i is computed by applying an activation function $f^{(1)}$ (for example the logistic function) to the weighted inputs and its bias term $b_i^{(1)}$:

$$h_i = f^{(1)}\left(b_i^{(1)} + \sum_{j=1}^{n} \mathbf{W}_{ij} x_j \right) \quad (3.5)$$

where $\mathbf{W}$ represents a weight matrix in which $\mathbf{W}_{ij}$ denotes the weight connecting input j to hidden neuron i. For the analysis conducted in this chapter, we make a binary prediction; hence, for the activation function in the output layer, we use the logistic (sigmoid) activation function, $f^{(2)}(x) = \dfrac{1}{1+e^{-x}}$ to obtain a response probability:

$$\pi = f^{(2)}\left(b^{(2)} + \sum_{j=1}^{n_h} \mathbf{v}_j h_j \right) \quad (3.6)$$

with n_h the number of hidden neurons and $\mathbf{V}$ the weight vector where $\mathbf{V}_j$ represents the weight connecting hidden neuron j to the output neuron. Examples of other commonly used transfer functions are the hyperbolic tangent $f(x) = \dfrac{e^x - e^{-x}}{e^x + e^{-x}}$ and the linear transfer function $f(x) = x$.

During model estimation, the weights of the network are first randomly initialized and then iteratively adjusted so as to minimize an objective function, for example, the sum of squared errors (possibly accompanied by a regularization term to prevent over-fitting). This iterative procedure can be based on simple gradient descent learning or more sophisticated optimization methods such as Levenberg-Marquardt or Quasi-Newton. The number of hidden neurons can be determined through a grid search based on validation set performance.

This approach can be formulated within SAS Enterprise Miner using the **Neural Network node** (Figure 3.3) within the Model tab.

Figure 3.3: Neural Network Node

It is worth noting that although Neural Networks are not necessarily appropriate for predicting PD under the Basel regulations, due to the model's non-linear interactions between the independent variables (customer attributes) and dependent (PD), there is merit in using them in a two-stage approach as discussed later in this chapter. They can also form a sense-check for an analyst in determining whether non-linear interactions do exist within the data so that these can be adjusted for in a more traditional logistic regression model. This may involve transforming an input variable by, for example, taking the log of an input or binning the input using a weights of evidence (WOE) approach. Analysts using Enterprise Miner can utilize the **Transform Variables node** to select the best transformation strategy and the **Interactive Grouping node** for selecting the optimal WOE split points.

3.2.4 Decision Trees

Classification and regression trees are decision tree models for categorical or continuous dependent variables, respectively, that recursively partition the original learning sample into smaller subsamples and reduces an impurity criterion $i()$ for the resulting node segments (Breiman, *et al* 1984). To grow the tree, one typically uses a greedy algorithm (which attempts to solve a problem by making locally optimal choices at each stage of the tree in order to find an overall global optimum) that evaluates a large set of candidate variable splits at each node t so as to find the 'best' split, or the split S that maximizes the weighted decrease in impurity:

$$\Delta i(s,t) = i(t) - p_L i(t_L) - p_R i(t_R) \quad (3.7)$$

where p_L and p_R denote the proportions of observations associated with node t that are sent to the left child node t_L or right child node t_R, respectively. A decision tree consists of internal nodes that specify tests on individual input variables or attributes that split the data into smaller subsets, as well as a series of leaf nodes assigning a class to each of the observations in the resulting segments. For Chapter 4, we chose the popular decision tree classifier C4.5, which builds decision trees using the concept of information entropy (Quinlan, 1993). The entropy of a sample S of classified observations is given by:

$$Entropy(S) = -p_1 \log_2(p_1) - p_0 \log_2(p_0) \quad (3.8)$$

where p_1 and p_0 are the proportions of the class values 1 and 0 in the sample S, respectively. C4.5 examines the normalized information gain (entropy difference) that results from choosing an attribute for splitting the data. The attribute with the highest normalized information gain is the one used to make the decision. The algorithm then recurs on the smaller subsets.

This approach can be formulated within SAS Enterprise Miner using the **Decision Tree node** (Figure 3.4) within the Model tab.

Figure 3.4: Decision Tree Node

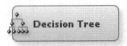

Analysts can automatically and interactively grow trees within the **Decision Tree node** of Enterprise Miner. An interactive approach allows for greater control over both which variables are split on and which split points are utilized.

Decision trees have many applications within credit risk, both within application and behavioral scoring as well as for collections and recoveries. They have become more prevalent in their usage within credit risk, particularly due to their visual natural and their ability to empirically represent how a decision was made. This makes internally and externally explaining complex logic easier to achieve.

3.2.5 Memory Based Reasoning

The k-nearest neighbor's algorithm (k-NN) classifies a data point by taking a majority vote of its k most similar data points (Hastie, *et al* 2001). The similarity measure used in this chapter is the Euclidean distance between the two points:

$$d(\mathbf{x}_i, \mathbf{x}_j) = \left\| \mathbf{x}_i - \mathbf{x}_j \right\| = \left[\left(\mathbf{x}_i - \mathbf{x}_j \right)^T \left(\mathbf{x}_i - \mathbf{x}_j \right) \right]^{1/2} \quad (3.9)$$

One of the major disadvantages of the k-nearest neighbor classifier is the large requirement on computing power. To classify an object, the distance between it and all the objects in the training set has to be calculated. Furthermore, when many irrelevant attributes are present, the classification performance may degrade when observations have distant values for these attributes (Baesens, 2003a).

This approach can be formulated within SAS Enterprise Miner using the **Memory Based Reasoning node** (Figure 3.5) within the Model tab.

Figure 3.5: Memory Based Reasoning Node

3.2.6 Random Forests

Random forests are defined as a group of un-pruned classification or regression trees, trained on bootstrap samples of the training data using random feature selection in the process of tree generation. After a large number of trees have been generated, each tree votes for the most popular class. These tree-voting procedures are collectively defined as random forests. A more detailed explanation of how to train a random forest can be found in Breiman (2001). For the Random Forests classification technique, two parameters require tuning. These are the number of trees and the number of attributes used to grow each tree.

The two meta-parameters that can be set for the Random Forests classification technique are the number of trees in the forest and the number of attributes (features) used to grow each tree. In the typical construction of a tree, the training set is randomly sampled, then a random number of attributes is chosen with the attribute with the most information gain comprising each node. The tree is then grown until no more nodes can be created due to information loss.

This approach can be formulated within SAS Enterprise Miner using the **HP Forest node** (Figure 3.6) within the HPDM tab (Figure 3.7).

Figure 3.6: Random Forest Node

Figure 3.7: Random Forest Node Location

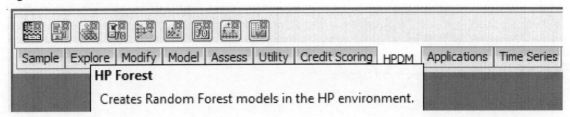

More information on the High-Performance Data Mining (HPDM) nodes within SAS Enterprise Miner can be found on the http://www.sas.com/ website by searching for "SAS High-Performance Data Mining".

3.2.7 Gradient Boosting

Gradient boosting (Friedman, 2001, 2002) is an ensemble algorithm that improves the accuracy of a predictive function through incremental minimization of the error term. After the initial base learner (most commonly a tree) is grown, each tree in the series is fit to the so-called "pseudo residuals" of the prediction from the earlier trees with the purpose of reducing the error. The estimated probabilities are adjusted by weight estimates, and the weight estimates are increased when the previous model misclassified a response. This leads to the following model:

$$F(\mathbf{x}) = G_0 + \beta_1 T_1(\mathbf{x}) + \beta_2 T_2(\mathbf{x}) + \ldots + \beta_u T_u(\mathbf{x}) \quad (3.10)$$

where G_0 equals the first value for the series, $T_1, \ldots, T_u$ are the trees fitted to the pseudo-residuals, and β_i are coefficients for the respective tree nodes computed by the Gradient Boosting algorithm. A more detailed explanation of gradient boosting can be found in Friedman (2001) and Friedman (2002). The meta-parameters which require tuning for a Gradient Boosting classifier are the number of iterations and the maximum branch used in the splitting rule. The number of iterations specifies the number of terms in the boosting series; for a binary target, the number of iterations determines the number of trees. The maximum branch parameter determines the maximum number of branches that the splitting rule produces from one node, a suitable number for this parameter is 2, a binary split.

This approach can be formulated within SAS Enterprise Miner using the **Gradient Boosting node** (Figure 3.8) found within the Model tab.

Figure 3.8: Gradient Boosting Node

3.3 Model Development (Application Scorecards)

In determining whether or not a financial institution will lend money to an applicant, industry practice is to capture a number of specific application details such as age, income, and residential status. The purpose of capturing this applicant level information is to determine, based on the historical loans made in the past, whether or not a new applicant looks like those customers who are known to be good (non-defaulting) or those customers we know were bad (defaulting). The process of determining whether or not to accept a new customer can be achieved through an application scorecard. Application scoring models are based on all of the captured demographic information at application, which is then enhanced with other information such as credit bureau scores or other external factors. Application scorecards enable the prediction of the binary outcome of whether a customer will be good (non-defaulting) or bad (defaulting). Statistically they estimate the likelihood (the probability value) that a particular customer will default on obligations to the bank over a particular time period (usually, a year).

Application scoring can be viewed as a process that enables a bank or other financial institutions to make decisions on credit approvals and to define risk attributes of potential customers. Therefore, this means that by applying a prudent application scoring process the approval rate for credit applications can be optimized based on the level of risk (risk-appetite) determined by the business.

3.3.1 Motivation for Application Scorecards

The main motivation behind developing an application scorecard model is to reduce the overall risk exposure when lending money in the form of credit cards, personal loans, mortgages etc. to new customers. In order to make these informed decisions, organizations rely on predictive models to identify the important risk inputs related to historical known good/bad accounts.

Application scorecard models enable organizations to balance the acceptance of as many applicants as possible whilst also keeping the risk level as low as possible. By automating this process through assigning scorecard points to customer attributes (such as age or income) a consistent unbiased treatment of applicants can be achieved. There are a number of additional benefits organizations can realize from the development of statistical scorecards, including:

- More clarity and consistency in decision making;
- Improved communication with customers and improved customer service;
- Reduction in employees' time spent on manual intervention;
- Quicker decision making at point of application;
- Consistency in score points allocation for every customer displaying the same details;

In order for financial institutions to make these informed data-based decisions through the deployment of an application scorecard, they have the following two options:

1. Utilize a third-party generic scorecard:
   a. Although an easier approach, this does not rely on in-house development; thus because it is not based on an organizations' own data, it may not provide the required level of accuracy;
   b. These can also be costly and do not allow for the development of in-house expertise.
2. Develop an in-house application scorecard predictive model to calculate a PD value for each customer:
   a. This method is more accurate and relevant to the organization, however, it does rely on the organization already holding sufficient historical information;
   b. Organizations can decide on the modeling approach applied, such as logistic regression, decision trees, or ensemble approaches (Section 3.2). This enables intellectual property (IP) to be generated internally and ensures complete control over the development process.

For those organizations subscribed to the internal ratings-based approach (IRB), either foundation or advanced, a key requirement is to calculate internal values for PD. As such, this chapter explores the approach an analyst would need in order to build a statistical model for calculating PD, and how this can be achieved through the use of SAS Enterprise Miner and SAS/STAT.

3.3.2 Developing a PD Model for Application Scoring

Accompanying this section is a full step-by-step tutorial on developing a PD model for application scoring, which is located in the tutorial section of this book. It is suggested that readers review the methodologies presented here and then apply their learnings through the practical steps given in the tutorial.

3.3.2.1 Overview

Typically large financial institutions want to create the most accurate and stable model possible based on a large database of known good/bad history. In order to create an application scoring model, an analyst must create a statistical model for calculating the probability of default value. The first step in creating an application scoring model is determining the time frame and the input variables from which the model will be built. You must prepare the data that will be used to create the application scoring model and identify the outcome time frame for which the model will be developed. The timeline in Figure 3.9 shows the outcome period for an application scoring model.

Figure 3.9: Outcome Time Frame for an Application Scoring Model

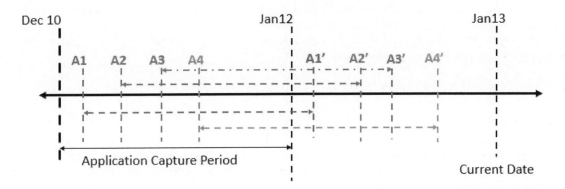

For example, A1 – A1' indicates the outcome period from which decisions can be made, given A1 as the date at which applicant 1 applied. During the performance period, a specific individual application account should be checked. Depending on whether it fits the bad definition defined internally, the target variable can be populated as either 1 or 0, with 1 denoting a default and 0 denoting a non-default.

All applications within the application capture period can then be considered for development in the model, which is commonly referred to as the Known Good/Bad (KGB) data sample. An observation point is defined as the date at which the status of account is classified as a good or bad observation. In the example depicted in Figure 3.9, observation point A1' is the classification date for application 1.

The independent variables that can be used in the prediction of whether an account will remain good or turn bad include such information as demographics relating to the applicant, external credit reference agency data, and details pertaining to previous performance on other accounts.

When deciding on a development time period, it is important to consider a time frame where a relatively stable bad rate has been observed. The reason for this is to minimize external fluctuations and ensure the robustness of future model performance. One of the major assumptions that we make during the development phase is that the future will reflect the past, and it is an analyst's responsibility to ensure that there are no abnormal time periods in the development sample. Abnormal time periods can be further adjusted for as part of a stress testing exercise, discussed further in Chapter 6.

3.3.2.2 Input Variables

As discussed in the previous section, a number of input variables captured at the point of application are assessed in the development of a predictive PD application scorecard model. These include information such as socio-demographic factors, employment status, whether it is a joint or single application, as well as other transactional level account information if an applicant already holds an account or line of credit. A data set can be constructed by populating this variable information for each application received and utilizing it as independent inputs into the predictive model. It is important to remember that only information known to us at the point of application can be considered in the development of an application scorecard, as the purpose of the model is to generalize to new customers wishing to take out credit where limited information is known.

3.3.2.3 Data Preparation

In determining the data that can be assessed in the creation of an application model, analysts must decide on which accounts can be considered. For a traditional application scorecard model, accounts that have historically been scored in a normal day-to-day credit offering process should be used. Accounts where an override has been applied or unusual or fraudulent activity has been identified should be excluded during the data preparation phase, as these can bias the final model.

Another important consideration in the data pooling phase is the length of observable history available on an account. This can vary between different lines of credit offered, with a typical time period of 18-24 months required for credit card accounts and up to 3-5 years for mortgage accounts.

With regards to determining whether qualifying accounts are defined as good or bad, a bad definition is usually decided upon based on the experience of the business. The bad definition itself should be aligned with the objectives of the financial organization. Depending on the case or product offered, the definition of bad can be as simple as whether a write-off has occurred, to determining the nature of delinquency over a fixed period of time. An analysis of the underlying data itself should also be undertaken in determining bad definitions for more complex products, such as those revolving lines of credit.

Typical examples of bad definitions are as follows:

- 90 days delinquent – this is defined to have occurred where a customer has failed to make a payment for 90-days consecutively (within the observation period).

- 2 x 30 days, or 2 x 60 days, or 1 x 90 days – this is defined to have occurred where a customer has been either 30 days delinquent twice, 60 days delinquent twice, or 90 days delinquent once (within the observation period).

Accounts falling between a bad or a good classification are defined as *indeterminates*. These arise where insufficient performance history has been obtained in order to make a classification. Analysts should decide on how best to treat these accounts and/or whether they should be excluded from the analysis. In terms of good accounts, as with bad, the definition should align to the objectives of the financial institution.

A typical definition for a good customer is:

- Where a customer has never gone into delinquency during a defined observation period. For example, if 90 days delinquent is defined as bad, then anything less than 90 days delinquent is defined as good.

Once the definitions have been decided upon, a binary good/bad target variable can then be populated. Using this definition rule, current accounts can then be determined as good or bad and the target flag can be appended to the modeling table for predictive modeling.

3.3.2.4 Model Creation Process Flow

Once a time frame for consideration has been determined, the independent input variables have been selected, and a definition for bad/default has been decided upon, a predictive model can be built. SAS Enterprise Miner provides analysts with a comprehensive selection of data mining and predictive modeling tools, a number of which can be directly used for application scorecard building. The SEMMA (Sample, Explore, Modify, Model, and Assess tabs) methodology should be employed as a guideline to the end-to-end model development process with three additional nodes available in the Credit Scoring tab. (Tutorial A at the end of this book details how to start a project within Enterprise Miner, with Tutorial B detailing the steps to develop the process flow for an Application Scorecard).

The following diagram (Figure 3.10) shows the model creation process flow for the accompanying sample application PD model. The subsequent sections describe nodes in the process flow diagram.

Figure 3.10: Application Scorecard Model Flow

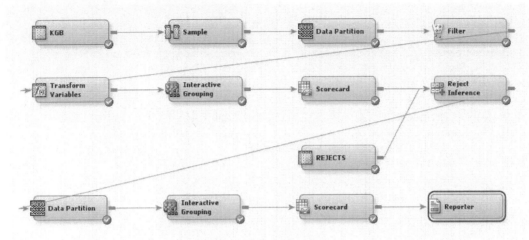

3.3.2.5 Known Good Bad Data

The Known Good Bad (KGB) data set in Figure 3.10 is an unbalanced sample consisting of only accepted applicants. In the KGB sample, there are 45,000 good and 1,500 bad accounts. A frequency weight of 30 Goods to 1 Bad has been applied in the data to balance the good observations with the bad observations. Bad has been defined as 90 days delinquent as per the definition in Section 3.3.2.3. Those observations not meeting the bad definition are considered good, so there are no indeterminate or rejected applicants (we explain how to incorporate the rejected candidates later in this chapter). In the KGB data set, the variable indicating whether an account is good or bad is called **GB**.

Figure 3.11: KGB Ratio of Goods to Bads

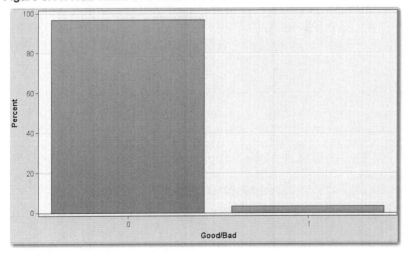

In Figure 3.11, **0** denotes a good customer and **1** denotes a bad customer.

3.3.2.6 Data Sampling

Traditionally, sampling has been applied in data mining processes where a larger database exists and it is either infeasible and/or too time-consuming to develop a model on the full data. To circumvent this, data miners and statisticians often randomly sample a larger volume of data down to a more manageable and computationally feasible size. In credit risk modeling, the number of defaults is also usually significantly lower than the population of non-defaults. As this can cause inefficiencies in the model development process, it is often the case to prepare a stratified data sample using a user-defined fixed proportion of defaulters to non-defaulters.

Where the required data does contain a disproportionate amount of defaulters to non-defaulters, an oversampling or biased stratified sampling methodology should be applied. In essence, the application of the oversampling technique biases the sampling to increase the number of rare-events (defaults) relative to the non-events (non-defaulters). Thus, each defaulter's record is duplicated multiple times to level the proportion of events to non-events.

To achieve this in SAS Enterprise Miner, you can make use of the **Sample node** (Sample tab) to perform both over and under sampling. Attach a **Sample node** to your input data, and change the properties to Sample method = Stratify, Stratified Criterion = Equal, and in the Train Variables property, select the Sample Role of stratification for your target variable.

In the sample provided within this book, oversampling has been applied to boost the event population so that the good and bad population each contribute to 50% of the data as shown in the following Figure 3.12.

Figure 3.12: KGB Reweighting of Good to Bads

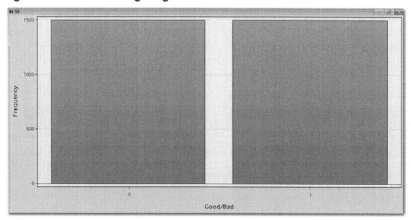

3.3.2.7 Outlier Detection and Filtering

As discussed in Chapter 2, an important stage to the modeling process is determining any input values that contain values outside of a reasonable range. Analysts should review the input data and determine whether outlying values exist. The **Filter node** in the Sample Tab can be utilized to filter outlying values from the data to produce a better model with more stable parameter estimates.

3.3.2.8 Data Partitioning

When a suitable volume of data observations are available, the best-practice approach is to further partition the data into development and hold-out samples. The model development set is then only utilized for model fitting with the hold-out validation sample utilized to obtain a final generalized estimate. The reason for this approach is to prevent the model from over-fitting and to increase the stability and life of the model. Typically, a split of 2/3rds or 70% training is used against a 1/3rd or 30% validation sample. Random sampling should be applied to achieve unbiased samples of the data. The **Data Partition node** (Sample Tab) in Enterprise Miner can be utilized for this task.

3.3.2.9 Transforming Input Variables

Prior to model development, it is often appropriate to create new variables or transform existing variables from the input data available for analysis. If implemented correctly, the new variables created can have a greater discrimination power in predicting the target outcome than the original raw variables had. Along with a prudent variable selection strategy, a large variety of variables can be created for consideration and then judged on their relative worth and predictive power in determining the dependent variable. Typical types of transformation that analysts may want to consider are aggregated values (means, ratios, proportions, and trends), distribution transformations (square, log, and exponential) or standardization/normalization of the variable values.

Once the appropriate derived variables have been created, they can then be utilized as inputs in the development of the predictive model. The **Transform Variables node** (Modify tab) can be utilized for this task. By setting the default method for the interval inputs to **Best,** the **Transform Variables node** will calculate all the appropriate transformations and select the best transformation based on the new r-square of the transformed input. You can also compare these r-square values to determine how much of an improvement has been attained over and above the original input variables.

3.3.2.10 Variable Classing and Selection

The process of variable classing and selection is utilized to either automatically or interactively group and select input variables relative to the event rate. The purpose of this is to improve the predictive power of characteristics by limiting the effect of outliers, managing the number of nominal attributes in a model, selecting the most predictive characteristics, and by varying an attributes Weight of Evidence (WOE) to smooth the points in the scorecard more linearly across the attributes. The number of associated points a characteristic's attribute is worth in a scorecard is determined based upon the risk of an attribute relative to all the other attributes for the same characteristic and the relative contribution of a characteristic to the overall score. The value for the relative risk of an attribute is determined by the calculated WOE amount as produced by the **Interactive Grouping node**. The **Interactive Grouping node** is a powerful tool in an analyst's armory for determining the optimal split points for interval variables in the calculation of the attribute values. In Figure 3.13, we can see that the number of unique levels of the variables AGE and INCOME are reduced to five levels. The WOE and grouping of each attribute is calculated and can then be used as an input into the scorecard model.

Figure 3.13: Interactive Grouping Node Report

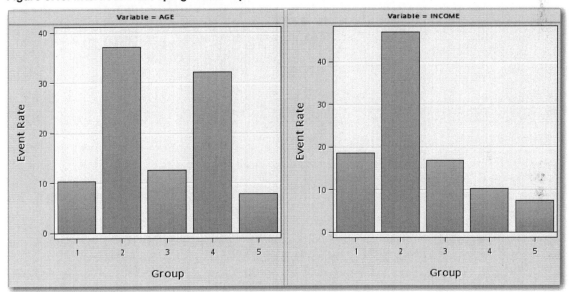

In practice, it is commonplace to utilize the WOE value for each attribute as a transformed input instead of using the group or the original continuous value. This allows for greater stability in the model and also a more explainable relationship between the relative risk of an input and the target. The **Interactive Grouping node** within the credit scoring tab of Enterprise Miner can be utilized for this task.

3.3.2.11 Modeling and Scaling

Once variable classing and selection has been attained, a logistic regression model can be applied using the WOE values as input into the prediction of the good/bad target flag. To apply a logistic regression model, analysts can utilize the **Scorecard node** or **Regression node** in Enterprise Miner or **proc logistic** in SAS/STAT. From the application of a regression model, coefficient estimates are determined for each of the variables that are present in the model. From these coefficient values, score points for each attribute can be scaled and estimated. The total score for an application is proportional to the log of the predicted good or bad odds for that application. It is general practice, however, in the presentation and reporting of these score point values, to scale them linearly and round them to the nearest integer value.

To implement this scaling, analysts can utilize the **Scorecard node**, where the scaling rule can easily be parameterized. The result of the scorecard points scaled for each attribute can then be exported as either a table or as a scorecard report. In Figure 3.14, each attribute for each characteristic contains different values for the score points relative to their risk in the prediction of the good/bad flag.

Figure 3.14: Example Scorecard Output

Scorecard		Scorecard Points
Age	AGE< 22	5
	22<= AGE< 28	13
	28<= AGE< 31	25
	31<= AGE< 47, _MISSING_	33
	47<= AGE	46
Credit Cards	CHEQUE CARD, MASTERCARD/EUROC, OTHER CREDIT CAR	47
	AMERICAN EXPRESS, NO CREDIT CARDS, VISA MYBANK, VISA OTHERS, _MISSING_, _UNKNOWN_	19
EC_card holders	0.00, _MISSING_, _UNKNOWN_	29
	1.00	19
Income	INCOME< 1500, _MISSING_	32
	1500<= INCOME< 2500	21
	2500<= INCOME< 3000	26
	3000<= INCOME< 4000	31
	4000<= INCOME	26

From the standard output, the smaller the value of score points assigned to an attribute, the higher the PD defined for that time period. However, this can also be reversed. For example, by analyzing Figure 3.14, we can see that an applicant whose age is less than 22 receives only 5 points (higher risk) whereas an applicant over the age of 47 receives 46 points (lower risk).

When the risk-related score points have been calculated, it is then possible to automate the process of accepting or rejecting a candidate at the point of application by determining a criteria for the number of points they must achieve. A binary Yes/No outcome or the total score points amount can then be provided back to the applicant dependent on business needs. If a Yes/No outcome is to be determined, a cutoff value must be defined that determines whether a No response or a Yes response is given. Program 3.2 depicts an example translation of the rule code derived from a scorecard, where accounts falling below a defined cutoff are classified as REJECT.

Program 3.2: Acceptance Logic

```
if score_points <= cutoff then Application="REJECT";
else Application="ACCEPT";
```

A way of determining this cutoff value is to analyze the Kolmogorov-Smirnov Plot, which is automatically produced in the **Scorecard node**. This plot can be found under **View ▶ Strength Statistics ▶ Kolmogorov-Smirnov Plot** (Figure 3.15).

Figure 3.15: Location of KS Plot

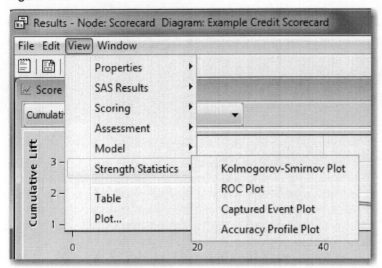

The peak of the KS Curve indicates the best discrimination between good and bad accounts. In the following Figure 3.16, this would equate to a cutoff value of approximately 180 points.

Figure 3.16: KS Plot

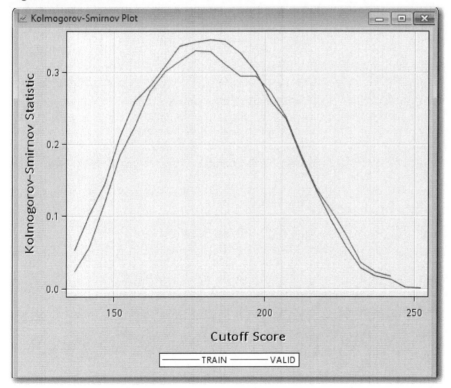

In Section 3.2, we discussed a number of classification techniques that can be considered in the prediction of those customers that are good and those that are bad. The accompanying KGB data set has been utilized in the following Figure 3.17 to display a potential setup that benchmarks each of these techniques against a typical application scorecard.

Figure 3.17: Scorecard Model Comparison Flow

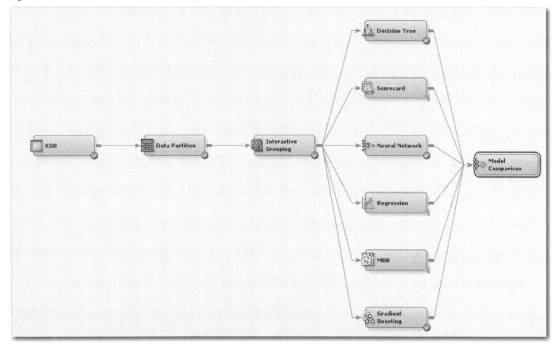

This is a first pass approach to an application scorecard prior to conducting any reject inference, which is discussed further in the next section. As we can see from the results table in Figure 3.18, the technique which offers the best discrimination in terms of ROC Index (AUC) on the validation set is a Neural Network, closer followed by a **Scorecard node**. A **Regression node** has also been shown, giving the same predictive power as the **Scorecard node**. This is to be expected, as the same logistic regression model underpins both the **Scorecard node** and the **Regression node**.

Figure 3.18: Scorecard Model Comparison Metrics

Selected Model	Predecessor Node	Model Node	Model Description	Target Variable	Target Label	Selection Criterion: Valid: Roc Index
Y	Neural	Neural	Neural Net...	GB	Good/Bad	0.712
	Scorecard	Scorecard	Scorecard	GB	Good/Bad	0.707
	Reg	Reg	Regression	GB	Good/Bad	0.707
	Tree	Tree	Decision Tr...	GB	Good/Bad	0.69
	MBR	MBR	MBR	GB	Good/Bad	0.672
	Boost	Boost	Gradient Bo...	GB	Good/Bad	0.651

Although a Neural Network does perform marginally better than a **Scorecard node,** the trade-off between a more black-box technique such as a Neural Network and the clarity of a Scorecard output would mean a preference for the Scorecard model. This does, however, give an example as to improvements that could be made in the Scorecard model if non-linear interactions between the input variables were analyzed further.

3.3.2.12 Reject Inference

Credit scoring models are built with a fundamental bias (selection bias). The sample data that is used to develop a credit scoring model is structurally different from the "through-the-door" population to which the credit scoring model is applied. The non-event or event target variable that is created for the credit scoring model is based on the records of applicants who were all accepted for credit. However, the population to which the credit scoring model is applied includes applicants who would have been rejected under the scoring rules that were used to generate the initial model.

One remedy for this selection bias is to use reject inference. The reject inference approach uses the model that was trained using the accepted applications to score the rejected applications. The observations in the rejected data set are then classified as either an inferred event (bad) or an inferred non-event (good). The inferred observations are then added to the KGB data set to form an augmented data set (AGB).

This augmented data set, which represents the "through-the-door" population, serves as the training data set for a second scorecard model. In the case of an application scorecard model, this secondary scorecard serves to recalibrate the regression coefficients to take the inferred rejected candidates into account.

To classify the rejected applicants data set, three methodologies are available in the **Reject Inference node**:

1. *Hard Cutoff Method* – the hard cutoff approach classifies observations as "good" or "bad" observations based on a cutoff score. If you choose hard cutoff as your inference method, you must specify a cutoff score in the Hard Cutoff properties. Any score below the hard cutoff value is allocated a status of "bad." You must also specify the rejection rate in General properties. The rejection rate is applied to the REJECTS data set as a frequency variable.

2. *Fuzzy Augmentation* – this approach uses partial classifications of "good" and "bad" to classify the rejects in the augmented data set. Instead of classifying observations as "good" and "bad," fuzzy classification allocates weight to observations in the augmented data set. The weight reflects the observation's tendency to be good or bad. The partial classification information is based on the $p\left(good\right)$ and $p\left(bad\right)$ from the model built on the KGB for the REJECTS data set. Fuzzy classification multiplies the $p\left(good\right)$ and $p\left(bad\right)$ values that are calculated in the Accepts for the Rejects model by the user-specified Reject Rate parameter to form frequency variables. This results in two observations for each observation in the Rejects data. One observation has a frequency variable $\left(\text{Reject Rate} \times p(good)\right)$ and a target variable of 0, and the other has a frequency variable $\left(\text{Reject Rate} \times p(\text{bad})\right)$ and a target value of 1. Fuzzy is the default inference method.

3. *Parceling Method* – the parceling methodology distributes binned scored rejects into "good" and bad" based on expected bad rates, $p\left(bad\right)$ that are calculated from the scores from the logistic regression model. The parameters that must be defined for parceling vary according to the Score Range method that you select in the Parceling Settings section. All parceling classifications require the Reject Rate setting as well as bucketing, score range, and event rate increase.

Although three methodologies are provided in Enterprise Miner, the industry standard is to use a reject weight which can be calculated using the fuzzy augmentation method, instead of using the hard cutoff or parceling methods for classifying the rejected applicants. The reasoning behind applying a fuzzy augmentation methodology is to apply a weighting value that reflects the tendency for an account to be good or bad, rather than an arbitrary split given by the hard cutoff approach. The eventual decision for determining the cutoff for either accepting or rejecting an application will be based upon a cost-revenue trade-off. This trade-off is finally taken into consideration only on the second and final scorecard, after reject inference has been applied.

The augmented good bad (AGB) data set is then re-partitioned using another **Data Partition node,** then a second model development flow is applied. At this point, classification still requires some modification, so a second **Interactive Grouping node** is added. Regression coefficients and score points are determined, the model is validated, and scorecard management reports are produced in the second **Scorecard node**.

3.3.2.13 Model Validation

Once the final application scorecard has been built and the input variables have been selected on the development sample, the model must be validated against an out-of-sample validation set. As discussed, typically this would be the 30% or 1/3rd hold-out validation sample determined in the data partition phase, but this could also be achieved by utilizing another data source. Validation qualifies the strength of the development sample and ensures that the model has not been over-fitted and can generalize well to unseen data. A simple visual way to validate a model is to overlay performance metrics such as the ROC or Score rankings plot with both the training output and validation output. Any large divergence observed between the overlaid performance metrics would be indicative of model over-fitting. A goodness-of-fit method such as the Chi-square test can also be employed to check empirically for divergence.

Figure 3.19 shows a score rankings overlay plot created by the **Scorecard node** in SAS Enterprise Miner across both the Train and Validation sets.

Figure 3.19: Score Rankings Overlay Plot

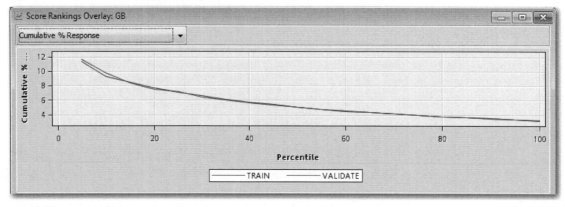

This figure demonstrates a strong fit for the cumulative percentage response rate across both the validation and training sets. Any divergence between the red and blue lines would indicate the model's difficulty generalizing to unseen data. To adjust for this, analysts can iteratively test different hold out samples using a cross validation approach. Cross validation can be applied in Enterprise Miner by utilizing the **Transform Variables node** and the **Start/End Groups nodes**. An example of this setup is presented in Figure 3.20:

Figure 3.20: Cross Validation Setup

The **Transform Variables node** is required to create a random segmentation ID for the k-fold groups' data to be used as cross validation indicators in the group processing loop. The formulation used to compute this is displayed in Figure 3.21.

Figure 3.21: Transformation Node Setup for Cross Validation

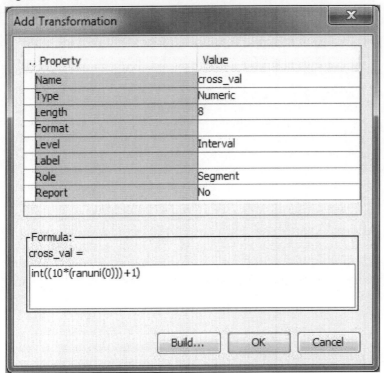

The above transformation creates a new variable called cross_val which contains 10 random segment ID values. Any number of cross validation folds can be created by adjusting the value in the formula. For example, for 5 folds, the formula becomes int((5*(ranuni(0)))+1). The property panel in the **Start Groups node** requires the General Mode to be Stratify. You need to set the cross_val in the Train Variables property panel to a Grouping Role of Stratification. Each model encapsulated within the **Start and End Group Processing nodes** will then be trained k times (k=10 in this example) using nine folds for training purposes and the remaining fold for evaluation (validation). A performance estimate for the classifier can then be determined by averaging the 10-validation estimates determined through the 10 runs of the cross validation.

Note: Typical experience through the implementation of an Enterprise Miner based scorecard build has resulted in a considerable reduction in scorecard develop time. In particular working with a large UK financial institution we were able to bring down a development process which had previously taken three-four months' worth of three analyst's time to two weeks of one analyst's time. This was predominately due to the complexity of the original process and the time it took for IT to implement any changes required, which was streamlined and simplified in EM, therefore minimizing re-work and reducing error. This level of time saving enables analysts to spend more time on additional analysis and adding more value to the business.

3.4 Model Development (Behavioral Scoring)

In the previous section, we looked at defining and calculating values for PD for application scorecards where customers are potentially new to the business. In this section, we look at how behavioral scorecards can be calculated and calibrated. We explore the considerations that must be made when applying a similar methodology to application scorecards to known accounts through behavioral scoring.

Behavioral scorecards are used in predicting whether an existing account will turn bad, or to vary current credit limits for a customer. The purpose here is to utilize behavioral information that has been captured for accounts over a period of time, such as credit utilization rates, months in arrears, overdraft usage, and to inform organizations as to the current risk state of accounts.

The calculation of values of PD in a behavioral scoring context is based upon an understanding of the transactional as well as demographic data available for analysis for accounts and customers. The purpose of creating behavioral scoring models is to predict if an existing customer/account is likely to default on its credit obligations. This enables financial institutions to constantly monitor and assess their customers/accounts and create a risk profile for each of these.

3.4.1 Motivation for Behavioral Scorecards

The motivation behind developing a PD model for behavioral scoring is to categorize current customers into varying risk groups based upon their displayed behavior with the credit lines they hold. This is a continual process once a customer has been offered credit and helps financial institutions to monitor the current risks related to the portfolios of credit they offer and to estimate their capital adequacy ratios. As discussed in the Introduction chapter, the major importance of these continual risk calculations are the relation of the current known risk of a portfolio of customers and the estimated risk weighted assets (RWA) based on these calculations. Calculating behavioral scores for customers can also identify early signs of high-risk accounts; in conjunction with remedial strategies, financial institutions can intervene before a prospective default occurs. In turn, this will aid the customer and also help reduce overall portfolio risk through the dynamic and effective management of portfolios. There are a number of other beneficial aspects to calculating behavioral scores, including:

- Faster reaction times to changes in market conditions affecting customer accounts;
- More accurate and timely decisions;
- Personalized credit limit setting for customer accounts;
- Standardized approach to policy implementation across portfolios;
- Improvements in the internal management of information systems (MIS)

The alternative to calculating behavioral scores for customers is to apply business-rule logic based on demographic and other customer attributes such as delinquency history or roll rate analysis. The main benefit of this approach is that it is relatively easy to implement and does not rely heavily on historical data analysis. However, it is limited by the experience of the decision maker and business rules can only be obtained from a smaller manageable size of variables, lacking the benefit of the full customer transactional information. By developing a statistical behavioral scoring model for estimating the PD of each customer, a more accurate and robust methodology can be embedded into an organization.

The following section 3.4.2 steps through an approach to apply a statistical behavioral scoring model to transactional customer accounts using SAS Enterprise Miner. We compare the differences between this approach and the variables utilized to application scorecard development.

3.4.2 Developing a PD Model for Behavioral Scoring

3.4.2.1 Overview

The creation of a behavioral scoring model requires the development of a predictive model given a certain time frame of events and a selection of input variables different to those detailed for application scoring models. Analysts must first determine a suitable time frame after an application has been accepted to monitor performance, gather defaulted occurrences, and calculate changes in input variables. An example time frame for sampling the data is shown in Figure 3.22 below:

Figure 3.22: Outcome Period for PD Model for Behavioral Scoring

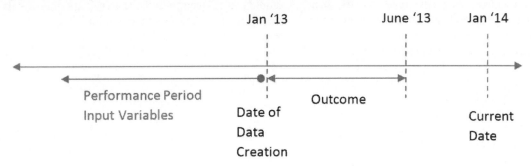

In the timeline above, January 2013 is the date of the creation of the data table, with the outcome period running from January to June 2013. All accounts that are active and non-defaulting as of the date of creation can be selected in the data table for modeling purposes. The behaviors of these accounts are then observed during the outcome window, with an appropriate target flag (good/bad) assigned to the accounts at the end of this period. Input variables can then be created based on the performance period prior to the date of data creation.

3.4.2.2 Input Variables

The input variables available in the development of a PD model for behavioral scoring are potentially more numerous than those used for application scorecard development due to the constant collection of data on active accounts. In order to determine current default behavior, a new set of risk characteristics needs to be identified based on the historical data for each account in a portfolio. The information typically collected at the account level includes aggregated summaries of transactions, utilizations of credit limits, and number of times in arrears over a defined period. All of these collected inputs form the independent variables into the predictive modeling process.

3.4.2.3 Data Preparation

In determining the data that can be assessed in the creation of a behavioral model, similar data preparation considerations should be made to those presented in the application scorecard section.

For a traditional behavioral scorecard model, accounts that have traditionally been scored in a normal day-to-day credit offering process should be used. Accounts where an override has been applied or unusual or fraudulent activity has been identified should be excluded during the data preparation phase, as these can bias the final model.

Another important consideration in the data pooling phase is the length of observable history available on an account. This can vary between different lines of credit offered, with a typical time period of 18-24 months required for credit card accounts and 3-5 years for mortgage accounts.

With regards to determining whether qualifying accounts are defined as good or bad, a bad definition is usually decided upon based on the experience of the business. The bad definition itself should be aligned with the objectives of the financial organization. Dependent on the case or product offered, the definition of bad can be as simple as whether a write-off has occurred, to determining the nature of delinquency over a fixed period of time. An analysis of the underlying data itself should also be undertaken to determine bad definitions for more complex products, such as revolving credit.

Typical examples of bad are as follows:

- 90 days delinquent – this is defined to have occurred where a customer has failed to make a payment for 90 days consecutively (within the observation period).

- 2 x 30 days, or 2 x 60 days, or 1 x 90 days – this is defined to have occurred where a customer has been either 30 days delinquent twice, 60 days delinquent twice, or 90 days delinquent once (within the observation period).

Accounts falling between a bad or a good classification are defined as indeterminates. These arise where insufficient performance history has been obtained in order to make a classification. Analysts should decide on how best to treat these accounts and/or whether they should be excluded from the analysis. In terms of good accounts as with bad, the definition should align to the objectives of the financial institution.

A typical definition for a good customer is:

- A customer who has never gone into delinquency during a defined observation period. For example, if 90 days delinquent is defined as bad, then anything less than 90 days delinquent is defined as good.

Once the definitions have been decided upon, a binary good/bad target variable can then be created based upon the defined definition. Using this definition rule, current accounts can then be determined as good or bad and the target flag can be appended to the modeling table for predictive modeling.

3.4.2.4 Model Creation Process Flow

Once a time frame for consideration has been determined, the independent input variables have been selected, and a definition for bad/default has been decided upon, a predictive model can be built. SAS Enterprise Miner provides analysts with a comprehensive selection of data mining and predictive modeling tools, a number of which can be used directly for behavioral scorecard building. The SEMMA (Sample, Explore, Modify, Model, and Assess) methodology should be employed as a guideline to the end-to-end model development process, with three additional nodes available in the Credit Scoring tab. Tutorial A at the end of this book details how to start a project within Enterprise Miner, with Tutorial B detailing the steps to develop the process flow for an Application Scorecard.

As we are applying a statistical methodology to the determination of a behavioral model, assumptions are made based on accurate, appropriate and complete data being used. Thus, the accuracy of the model created does rely on the quality of the data being used. SAS Enterprise Miner has a number of data mining techniques to test the adequacy of the data and to handle missing or outlier values. This subject is covered in comprehensive detail in Chapter 2. The techniques presented in Chapter 2 should be considered before the development of the predictive model is undertaken.

The development of an accurate predictive model for behavioral scoring requires the following key aspects to be in place:

- The structure of the underlying table should be in an Analytical Base Table (ABT) format, where one record is present per customer or account. This table contains the independent predictive measures plus the target good/bad flag.

- An understanding of the relationship between the input variables and their relation to the target should also be developed. This can be achieved through variable analysis or correlation analysis using either the **Variable Clustering node** in Enterprise Miner or **proc varclus** in SAS/STAT.

Once the understanding of the data has been developed, analysts can embark on the model process flow development in SAS Enterprise Miner. The following Figure 3.23 shows a typical model creation process flow for a PD model for behavioral scoring.

Figure 3.23: PD Model for Behavioral Scoring Flow

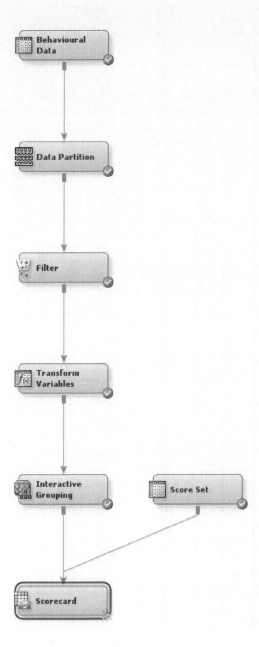

The development process for behavioral scorecards is similar to that of application scorecards; however, as you are identifying the risk characteristics of customers in flight as opposed to offering them credit prior, there is no requirement for a reject inference stage.

Once the final behavioral scoring model has been developed, the scoring logic can be exported using a **Score node** and applied back to the database of accounts. As transactional patterns change at the account level, the model can automatically be applied to adjust for the current level of risk associated to an account. Monitoring of the behavioral model once live is of paramount importance, and the topics of such will be discussed in the following sections.

As discussed above, the process for behavioral modeling is similar with regards to application modeling, however different behavioral independent variables can be considered. There is also a key difference in purpose; while an application model will be utilized to determine which customers will be offered credit, behavioral models are key in determining whether to offer a rise/decrease in credit limit, additional lines of credit, or a change in APR. These are all appropriate applications for current customers through their lifecycle.

3.5 PD Model Reporting

3.5.1 Overview

With the culmination of the development of an application or behavioral scoring model, a complete set of reports need to be created. Reports include information such as model performance measures, development score and scorecard characteristics distributions, expected bad or approval rate charts, and the effects of the scorecard on key subpopulations. Reports facilitate operational decisions such as deciding the scorecard cutoff, designing account acquisition and management strategies, and monitoring scorecards.

The majority of the model performance reports an analyst will be required to produce can be automatically generated using SAS Enterprise Miner, with further reporting available in both SAS Enterprise Guide and SAS Model Manager. A full detailing of the PD reports generated in Model Manager can be located in Chapter 7.

The key reporting parameters that should be considered during the creation of reports are detailed in the following sections.

3.5.2 Variable Worth Statistics

In determining the predictive power, or variable worth, of constituent input variables, it is common practice to calculate the information value (IV) or Gini Statistic based on their ability to separate the good/bad observations into two distinct groups.

After binning input variables using an entropy-based procedure implemented in SAS Enterprise Miner, the information value of a variable with k bins is given by:

$$IV = \sum_{i=1}^{k} \left[\left(\frac{n_1(i)}{N_1} - \frac{n_0(i)}{N_0} \right) \ln \left(\frac{n_1(i)/N_1}{n_0(i)/N_0} \right) \right] \quad (3.11)$$

where $n_0(i), n_1(i)$ denote the number of non-events (non-defaults) and events (defaults) in bin i, and N_0, N_1 are the total number of non-events and events in the data set, respectively.

This measure allows analysts not only to conduct a preliminary screening of the relative potential contribution of each variable in the prediction of the good/bad accounts, but also to report on the relative worth of each input in the final model.

Figure 3.24: Variable Worth Statistics

Variable	Label	Gini Statistic	Information Value
AGE	Age	32.94	0.379
TMJOB1	Time at Job	23.129	0.209
INCOME	Income	24.631	0.207
STATUS	Status	22.487	0.203
CARDS	Credit Cards	17.41	0.155
EC_CARD	EC_card holders	15.697	0.133
PERS_H	Num in Household	18.434	0.132
TEL	Telephone	9.164	0.058
PROF	Profession	9.659	0.056
CAR	Type of Vehicle	9.455	0.051
TMADD	Time at Address	10.933	0.044
NMBLOAN	Num Mybank Loans	9.026	0.041
CHILDREN	Num of Children	10.45	0.04
REGN	Region	8.984	0.028
CASH	Requested cash	7.922	0.025
PRODUCT	Type of Business	7.471	0.022
FINLOAN	Num finished Loans	6.434	0.017
LOANS	Num of running loans	5.35	0.015
DIV	Large region	5.025	0.013
BUREAU	Credit Bureau Risk Class	4.351	0.009
NAT	Nationality	2.302	0.004
TITLE	Title	2.928	0.004
RESID	Residence Type	0	0
LOCATION	Location of Credit Bureau	0	0

As we can see from the above Figure 3.24, the input variables that offer the highest information value and Gini Statistic are Age, Time at Job, and Income. Typically practitioners use the following criteria in assessing the usefulness of Information Value results:

- less than 0.02: unpredictive
- 0.02 to 0.1: weak
- 0.1 to 0.3: medium
- 0.3+: strong

In terms of relation to the Gini Statistic, this would loosely translate to:

- less than 8: unpredictive
- 8 to 15: weak
- 15 to 30: medium
- 30+: strong

Through the use of the **Interactive Grouping node** is SAS Enterprise Miner, analysts can define their own cutoff values for automatically rejected variables based upon either their IV or Gini Statistic. More information on the Gini and Information Value statistics can be found in the Enterprise Miner help section for **the Interactive Grouping node**.

3.5.3 Scorecard Strength

After evaluating the predictive power of the independent variables used in the model, the next step is to evaluate the strength of the model based on the scorecard produced. Within the project flow for an application scorecard (Figure 3.10) and behavioral scorecard (Figure 3.23), right-click the **Scorecard node** and select **Results…**. This will open a new window displaying the automatically created reporting output from the scorecard model. Double-click the **Fit Statistics** window to view the key performance metrics used in determining the value of a model such as the Gini Statistic, Area Under ROC (AUC), and K-S Statistic across both the development train sample and validation sample. To view plots of these metrics, navigate to **View -> Strength** Statistics, where an option is given to display the K-S Plot, ROC Plot, and Captured Event Plot.

- The *Kolmogorov-Smirnov Plot* shows the Kolmogorov-Smirnov statistics plotted against the score cutoff values. The Kolmogorov-Smirnov statistic measures the maximum vertical separation at a scorecard point between the cumulative distributions of applicants who have good scores and applicants who have bad scores. The Kolmogorov-Smirnov plot displays the Kolmogorov-Smirnov statistics over the complete cutoff score range. The peak of the K-S Plot is a good indication of where the cutoff in score points should be set in accepting new applicants.
- The *ROC Plot* (4.3.3) shows the measure of the predictive accuracy of a logistic regression model. It displays the Sensitivity (the true positive rate) versus 1-Specificity (the false positive rate) for a range of cutoff values. The best possible predictive model would fall in the upper left hand corner, representing no false negatives and no false positives.
- In a *Captured Event Plot*, observations are first sorted in ascending order by the values of score points. The observations are then grouped into deciles. The Captured Event plot displays the cumulative proportion of the total event count on the vertical axis. The horizontal axis represents the cumulative proportion of the population.

Analysts can use a combination of these metrics in determining an acceptable model for implementation and for deciding on the cutoff for applicant selection. Models demonstrating an AUC greater than 0.7 on the validation can be considered as predictive in terms of their discrimination of goods and bads. The Gini Coefficient can also be derived from the AUC by the formulation $Gini = 2AUC - 1$.

3.5.4 Model Performance Measures

The model performance measures that should be estimated under the Basel II back testing criteria include the Area Under the ROC Curve (AUC), Accuracy Ratio (Gini), Error Rate, K-S Statistic, Sensitivity, and Specificity, which are all automatically calculated by SAS Enterprise Miner. A number of other performance measures that analysts typically want to employ are automatically calculated by SAS Model Manager for model monitoring reports and are detailed in Chapter 7. These can be calculated manually from the output tables provided by Enterprise Miner, but would need to formulated in SAS/STAT code.

3.5.5 Tuning the Model

The model should be monitored continuously and refreshed when needed (based on specific events or degradation in model accuracy). For example, through the monitoring of the selected accuracy measure (Gini, ROC and/or Captured Event Rate) thresholds should be determined (10% drop in Gini) to signal the need to refresh the analytical model. This incorporates the process of recalculating the performance metrics on the current view of the data and verifying how these values change over a time period. If the value of the measure drops below the selected threshold value, the analytical model should be recalibrated. Model inputs can be monitored using automated reports generated by SAS Model Manager or manually coded using SAS Enterprise Guide. For more information about the model-input-monitoring reports, see Chapter 7.

3.6 Model Deployment

Once a model has been built, the next stage is to implement the resulting output. SAS Enterprise Miner speeds up this process by providing automatically generated score code output at each stage of the process. By appending a **Score node** (Figure 3.25) from the Assess tab, the optimized SAS scoring code, Java, PMML, or C code is parceled up for deployment.

Figure 3.25: Score Node

3.6.1 Creating a Model Package

SAS Enterprise Miner enables analysts to create model packages. Model packages enable you to:

- Document a data mining project,
- Share results with other analysts,
- Recreate a project at a later date.

After you run a modeling node, there are a number of ways to export the contents of your model to a model package:

1. Right-click the node and select **Create Model Package** from the list,
2. Click the node and select the **Create Model Package** action button,
3. Or, click the node and select **Actions => Create Model Package** from the main menu.

By clicking **Create Model Package** (Figure 3.26) you are prompted to enter a name for the model package. It is best practice to choose a name which meaningfully describes either the function or purpose of the process flow or modeling tool, for reporting purposes (for example PD_app_model_v1). By default, model packages are stored within the Reports subdirectory of your project directory. The folder is named by a combination of the name that you specified when you saved the model package and a string of random alphanumeric characters.

Figure 3.26: Creating a Model Package

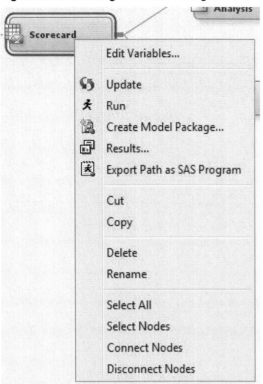

Model package folders contain the following files:

1. **miningresult.sas7bcat** — SAS catalog that contains a single SLIST file with metadata about the model
2. **miningResult.spk** — the model package SPK file
3. **miningResult.xml** — XML file that contains metadata about the modeling node. This file contains the same information as miningresult.sas7bcat

Right-clicking a model package in the project panel allows one to perform several tasks with the model package: open, delete, recreate the diagram, and save as another package.

3.6.2 Registering a Model Package

In the project panel, right-click the model package created and select **Register** (Figure 3.27). Click the **Description** tab and type in a description. Click the **Details** tab to show what additional metadata is saved with the model registration. Once a model is registered to Metadata, the model can be shared with any other data miners in the organization who have access to the model repository.

Figure 3.27: Registering a Model Package to Metadata

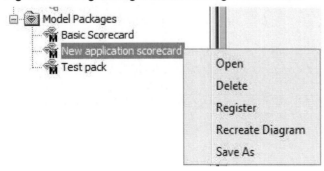

Once the model package has been registered in Enterprise Miner, the model can be used to score new data. One way of achieving this is through the use of the **Model Scoring** task in SAS Enterprise Guide (Figure 3.28).

Figure 3.28: Model Scoring in SAS Enterprise Guide

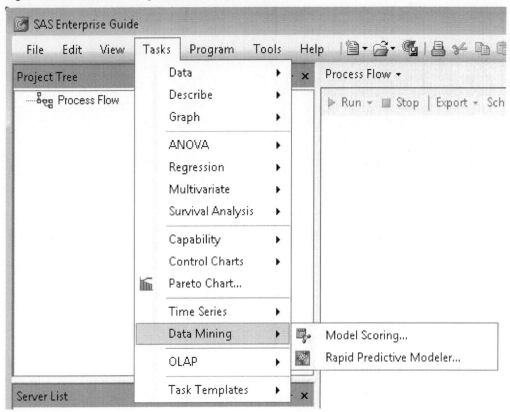

First, load the data you wish to score into Enterprise Guide, then click the Model Scoring Task under **Tasks => Data Mining.** In the scoring model screen, select **Browse,** and locate the folder the Model Package was registered to. Once you have mapped the input variables used in the model to the data you wish to score, run the task and an outputted scored set will be produced.

Once a model has been registered, it can also be called by other SAS applications such as SAS Model Manager or SAS Data Integration Studio. By accessing a registered model in SAS Model Manager, performance monitoring can be undertaken over the life of the model. Further information on the performance reports that can be generated in SAS Model Manager can be found in Chapter 7.

3.7 Chapter Summary

In this chapter, the processes and best practices for the development of a PD model for both application and behavioral scorecards using SAS Enterprise Miner have been given. We have looked at the varying types of classification techniques that can be utilized, as well as giving practical examples of the development of an Application and Behavioral Scorecard.

In the following chapter, we focus on the development of Loss Given Default (LGD) models and the considerations with regard to the distribution of LGD that have to be made for modeling this parameter. A variety of modeling approaches are discussed and compared in order to show how improvements over the traditional industry approach of linear regression can be made.

3.8 References and Further Reading

Altman, E.I. 1968. "Financial Ratios, Discriminant Analysis and the Prediction of Corporate Bankruptcy." The Journal of Finance, 23(4), 589-609.

Baesens, B. 2003a. "Developing intelligent systems for credit scoring using machine learning techniques." PhD Thesis, Faculty of Economics, KU Leuven.

Basel Committee on Banking Supervision. 2004. *International Convergence of Capital Measurement and Capital Standards: A Revised Framework.* Bank for International Settlements.

Bishop, C.M. 1995. *Neural Networks for Pattern Recognition.* Oxford University Press: Oxford, UK.

Bonfim, D. 2009. "Credit risk drivers: Evaluating the contribution of firm level information and of macroeconomic dynamics." Journal of Banking & Finance, 33(2), 281-299.

Breiman, L. 2001. "Random Forests." Machine Learning, 45(1), 5-32.

Breiman, L., Friedman, J., Stone, C., and Olshen, R. 1984. *Classification and Regression Trees.* Chapman & Hall/CRC.

Carling, K., Jacobson, T., Lindé J. and Roszbach, K. 2007. "Corporate Credit Risk Modeling and the Macroeconomy." Journal of Banking & Finance, 31(3), 845-868.

Fernandes, J.E. 2005. "Corporate credit risk modeling: Quantitative rating system and probability of default estimation," mimeo.

Friedman, J. 2001. "Greedy function approximation: A gradient boosting machine." The Annals of Statistics, 29(5), 1189-1232.

Friedman, J. 2002. "Stochastic gradient boosting." Computational Statistics & Data Analysis, 38(4), 367-378.

Giambona, F., and Iacono, V.L. 2008. "Survival models and credit scoring: some evidence from Italian Banking System." 8th International Business Research Conference, Dubai, 27th-28th March 2008

Güttler, A. and Liedtke, H.G. 2007. "Calibration of Internal Rating Systems: The Case of Dependent Default Events." Kredit und Kapital, 40(4), 527-552

Guettler, A and Liedtke, H.G. 2007. "Calibration of Internal Rating Systems: The Case of Dependent Default Events." Kredit und Kapital, 40, 527-552

Hastie, T., Tibshirani, R., and Friedman, J. 2001. *The Elements of Statistical Learning, Data Mining, Inference, and Prediction.* Springer: New York.

Hosmer, D.W., and Stanley, L. 2000. *Applied Logistic Regression,* 2nd ed. New York; Chichester, Wiley.

Kiefer, N.M. 2010. "Default Estimation and Expert Information." Journal of Business & Economic Statistics, 28(2), 320-328.

Martin, D. 1977. "Early warning of bank failure: A logit regression approach." Journal of Banking & Finance, 1(3), 249–276.

Miyake, M., and Inoue, H. 2009. "A Default Probability Estimation Model: An Application to Japanese Companies." Journal of Uncertain Systems, 3(3), 210–220

Ohlson, J. 1980. "Financial ratios and the probabilistic prediction of bankruptcy." Journal of Accounting Research, 109–131.

Quinlan, J.R. 1993. *C4.5 Programs for Machine Learning.* Morgan Kaufmann: San Mateo, CA.

Tarashev, N.A. 2008. "An Empirical Evaluation of Structural Credit-Risk Models." International Journal of Central Banking, 4(1), 1-53.

Tasche, D. 2003. "A Traffic Lights Approach to PD Validation," Frankfurt.

Walker, S.H., and Duncan, D.B. 1967. "Estimation of the Probability of an Event as a Function of Several Independent Variables." Biometrika, 54, 167-179

West, D. 2000. "Neural network credit scoring models." Computers & Operations Research, 27(11-12), 1131–1152.

Chapter 4 Development of a Loss Given Default (LGD) Model

4.1 Overview of Loss Given Default

Loss Given Default (LGD) is the estimated economic loss, expressed as a percentage of exposure, which will be incurred if an obligor goes into default (also referred to as 1 – the recovery rate). Producing robust and accurate estimates of potential losses is essential for the efficient allocation of capital within financial organizations for the pricing of credit derivatives and debt instruments (Jankowitsch et al., 2008). Banks are also in the position to gain a competitive advantage if an improvement can be made to their internally made loss-given default forecasts.

Whilst the modeling of probability of default (PD) has been the subject of many textbooks during the past few decades, literature detailing recovery rates has only emerged more recently. This increase in literature on recovery rates is due to the continued efforts by financial organizations in the implementation of the Basel Capital Accord.

In this chapter, a step-by-step process for the estimation of LGD is given, through the use of SAS/STAT techniques and SAS Enterprise Miner. At each stage, examples will be given using real world financial data. This chapter also demonstrates, through a case study, the development and computation of a series of competing models for predicting Loss Given Default to show the benefits of each modeling methodology. A full description of the data used within this chapter can be found in the appendix section of this book.

Although the focus of this chapter is on retail credit, it is worth noting that a clear distinction can be made between those models developed for retail credit and corporate credit facilities. As such, this section has been sub-divided into four categories distinguishing the LGD topics for retail credit, corporate credit, economic variables, and downturn LGD.

4.1.1 LGD Models for Retail Credit

Bellotti and Crook evaluate alternative regression methods to model LGD for credit card loans (2007). This work was conducted on a large sample of credit card loans in default, and also gives a cross-validation framework using several alternative performance measures. Their findings show that fractional logit regression gives the highest predictive accuracy in terms of mean absolute error (MAE). Another interesting finding is that simple OLS is as good, if not better, than estimating LGD with a Tobit or decision tree approach.

In Somers and Whittaker, quantile regression is applied in two credit risk assessment exercises, including the prediction of LGD for retail mortgages (2007). Their findings suggest that although quantile regression may be usefully applied to solve problems such as distribution forecasting, when estimating LGD, in terms of R-square, the model results are quite poor, ranging from 0.05 to a maximum of 0.2.

Grunert and Weber conduct analyses on the distribution of recovery rates and the impact of the quota of collateral, the creditworthiness of the borrower, the size of the company, and the intensity of the client relationship on the recovery rate (2008). Their findings show that a high quota of collateral leads to a higher recovery rate.

In Matuszyk et al., a decision tree approach is proposed for modeling the collection process with the use of real data from a UK financial institution (2010). Their findings suggest that a two-stage approach can be used to estimate the class a debtor is in and to estimate the LGD value in each class. A variety of regression models are provided with a weight of evidence (WOE) approach providing the highest R-square value.

In Hlawatsch and Reichling, two models for validating relative LGD and absolute losses are developed and presented, a proportional and a marginal decomposition model (2010). Real data from a bank is used in the testing of the models and in-sample and out-of-sample tests are used to test for robustness. Their findings suggest that both their models are applicable without the requirement for first calculating LGD ratings. This is beneficial, as LGD ratings are difficult to develop for retail portfolios because of their similar characteristics.

4.1.2 LGD Models for Corporate Credit

Although few studies have been conducted with the focus on forecasting recoveries, an important study by Moody's KMV gives a dynamic prediction model for LGD modeling called LossCalc (Gupton and Stein, 2005). In this model, over 3000 defaulted loans, bonds, and preferred stock observations occurring between the period of 1981 and 2004 are used. The LossCalc model presented is shown to do better than alternative models such as overall historical averages of LGD, and performs well in both out-of-sample and out-of-time predictions. This model allows practitioners to estimate corporate credit losses to a better degree of accuracy than was previously possible.

In the more recent literature on corporate credit, Acharya et al. use an extended set of data on U.S. defaulted firms between 1982 and 1999 to show that creditors of defaulted firms recover significantly lower amounts, in present-value terms, when their particular industry is in distress (2007). They find that not only an economic-downturn effect is present, but also a fire-sales effect, also identified by Shleifer and Vsihny (1992). This fire-sales effect means that creditors recover less if the surviving firms are illiquid. The main finding of this study is that industry conditions at the time of default are robust and economically important determinants of creditor recoveries.

An interesting study by Qi and Zhao shows the comparison of six statistical approaches to estimate LGD (including regression trees, neural networks and OLS with and without transformations) (2011). Their findings suggest that non-parametric methods such as neural networks outperform parametric methods such as OLS in terms of model fit and predictive accuracy. It is also shown that the observed values for LGD in the corporate default data set display a bi-modal distribution with focal points around 0 and 1. This paper is limited, however, by the use of a single corporate defaults data set of a relatively small size (3,751 observations). Extending this study over multiple data sets and including a variety of additional techniques would therefore add to the validity of the results.

4.1.3 Economic Variables for LGD Estimation

Altman finds that when the recovery rates are regressed on the aggregate default rate as an indicator of the aggregate supply of defaulted bonds, a negative relationship is given (2006). However, when macroeconomic variables such as GDP growth are added as additional explanatory variables, they exhibit low explanatory power for the recovery rates. This indicates that in the prediction of the LGD (recovery rate) at account level, macroeconomic variables do not add anything to the models which only incorporate individual loan-related variables derived from the data.

Hu and Perraudin present evidence that aggregate quarterly default rates and recovery rates are negatively correlated (2002). This is achieved through the use of Moody's historical bond market data from 1971-2000. Their conclusions suggest that recoveries tend to be low when default rates are high. It is also concluded that typical correlations for post-1982 quarters are -22%. Whereas, with respect to the full time period of 1971-2000, correlations are typically lower (-19%).

Caselli et al. verify the existence of a relation between the loss given default rate (LGDR) and macroeconomic variables (2008). Using a sizeable number of bank loans (11,649) concerning the Italian market, several models are tested in which LGD is expressed as a linear combination of the explanatory variables. They find that for households, LGDR is more sensitive to the default-to-loan-ratio, the unemployment rate, and household consumption. For small to medium enterprises (SMEs) however, LGDR is influenced to a great extent by the GDP growth rate and total number of people employed. The estimation of the model coefficients in this analysis was achieved by using a simple OLS regression model.

In an extension to their prior work, Bellotti and Crook (2009), add macroeconomic variables to their regression analysis for retail credit cards. The conclusions drawn indicate that although the data used has limitations in terms of the business cycle, adding bank interest rates and unemployment levels as macroeconomic variables into an LGD model yields a better model fit and that these variables are statistically significant explanatory variables.

4.1.4 Estimating Downturn LGD

Several studies have approached the problem of estimating downturn LGD from varying perspectives. For example, in Hartmann-Wendels and Honal, a linear regression model is implemented with the use of a dummy variable to represent downturn LGD (2006). The findings from this study show that downturn LGD exceeds default-weighted average LGD by eight percent. In Rosch and Scheule, alternative concepts for the calculation of downturn LGD are given on Hong Kong mortgage loan portfolios (2008). Their findings show that the empirical calibration of the downturn LGD agrees with regulatory capital adequacy. Their empirical analysis also highlights that the asset correlations given by the Basel Committee on Banking Supervision exceed the values empirically estimated, and therefore could lead to banks overproviding for capital (2006).

In addition to the papers discussed in this section, the following papers provide information on other areas of loss given default modeling: Benzschawel et al. (2011); Jacobs and Karagozoglu (2011); Sigrist and Stahel (2010); Luo and Shevchenko (2010); Bastos (2010); Hlawatsch and Ostrowski (2010); Li (2010); Chalupka and Kopecsni (2009). As discussed, the LGD parameter measures the economic loss, expressed as percentage of the exposure, in case of default. This parameter is a crucial input to the Basel II capital calculation as it enters the capital requirement formula in a linear way (unlike PD, which comparatively has a smaller effect on minimal capital requirements). Hence, changes in LGD directly affect the capital of a financial institution and thus, its long-term strategy. It is thus of crucial importance to have models that estimate LGD as accurately as possible.

This is not straightforward, however, as industry models typically show low R^2 values. Such models are often built using ordinary least squares regression or regression trees, even though prior research has shown that LGD typically displays a non-linear bi-modal distribution with spikes around 0 and 1 (Bellotti and Crook, 2007). In the literature, the majority of work to date has focused on the issues related to PD estimation, whereas only more recently, academic work has been conducted into the estimation of LGD (Bellotti and Crook, 2009, Loterman et al., 2009, Matuszyk et al., 2010).

In this chapter, as well as a step-by-step guide to the development of an LGD model in Enterprise Miner, a number of regression algorithms will use example LGD data with the aim of informing the reader with a better understanding of how each technique performs in the prediction of LGD. The regression models employed include one-stage models, such as those built by ordinary least squares, beta regression, artificial neural networks, and regression trees, as well as two-stage models which attempt to combine the benefits of multiple techniques. Their performances are determined through the calculation of several performance metrics, which are in turn meta-ranked to determine the most predictive regression algorithm. The performance metrics are again compared using Friedman's average rank test, and Nemenyi's post-hoc test is employed to test the significance of the differences in rank between individual regression algorithms. Finally, a variant of Demšar's significance diagrams will be plotted for each performance metric to visualize their results.

There has been much industry debate as to the best techniques to apply in the estimation of LGD, given its bi-modal distribution. The aim of the case study in this chapter is to detail the predictive power of commonly used techniques, such as linear regression, with transformations and compare them to more advanced machine learning techniques such as neural networks. This is to better inform industry practitioners as to the comparable ability of potential techniques and to add to the current literature on both the topics of loss given default and applications of domain specific regression algorithms.

4.2 Regression Techniques for LGD

Whereas in Chapter 3, we looked at potential classification techniques that can be applied in industry in the modeling of PD, in this section we detail the proposed regression techniques to be implemented in the modeling of LGD. The experiments comprise a selection of one-stage and two-stage techniques. One-stage techniques can be divided into linear and non-linear techniques. The linear techniques included in this chapter model the (original or transformed) dependent variable as a linear function of the independent variables, whereas non-linear techniques fit a non-linear model to the data set. Two-stage models are a combination of the aforementioned one-stage models. These either combine the comprehensibility of an OLS model with the added predictive power of a non-linear technique, or they use one model to first discriminate between zero-and-higher LGDs and a second model to estimate LGD for the subpopulation of non-zero LGDs.

A regression technique fits a model $y = f(\mathbf{x}) + e$ onto a data set, where y is the dependent variable, $\mathbf{X}$ is the independent variable (or variables), and e is the residual.

Table 4.1 details the regression techniques discussed in this chapter for the estimation of LGD:

Table 4.1: Regression Techniques Used for LGD modeling

Regression Techniques	Description
Linear	
1. Ordinary Least Squares (OLS)	Linear regression is the most common technique to find optimal parameters to fit a linear model to a data set (Draper and Smith, 1998). OLS estimation produces a linear regression model that minimizes the sum of squared residuals for the data set.
2. Ordinary Least Squares with Beta Transformation (B-OLS)	Before estimating an OLS model, B-OLS fits a beta distribution to the dependent variable (LGD) (Gupton and Stein, 2002). The purpose of this transformation is to better meet the OLS normality assumption.
3. Beta Regression (BR)	Beta regression uses maximum likelihood estimation to produce a generalized linear model variant that allows for a dependent variable that is beta-distributed conditional on the input variables (Smithson and Verkuilen, 2006).
4. Ordinary Least Squares with Box-Cox Transformation (BC-OLS)	Box-Cox transformation/OLS selects an instance of a family of power transformations to improve the normality of the dependent variable (Box and Cox, 1964).
Non-linear	
1. Regression Trees (RT)	Regression tree, sometimes referred to as classification and regression trees (CART), algorithms produce a decision tree for the dependent variable by recursively partitioning the input space based on a splitting criterion, such as weighted reduction in within-node variance (Breiman, et al. 1984).
2. Artificial Neural Networks (ANN)	Artificial Neural Networks produce an output value by feeding inputs through a network whose subsequent nodes apply some chosen activation function to a weighted sum of incoming values. The type of ANN considered in this chapter is the popular multilayer perceptron (MLP) (Bi and Bennet, 2003).
Log+(non-)linear	
1. Logistic regression + OLS, B-OLS, BR, BC-OLS, RT, or ANN	This class of two-stage (mixture) modeling approaches uses logistic regression to first estimate the probability of LGD ending up in the peak at 0 (LGD≤0) or to the right of it (LGD>0) (Matuszyk, et al. 2010). A second-stage non-linear regression model is built using only the observations for which LGD>0. An LGD estimate is then produced by weighting the average LGD in the peak and the estimate produced by the second-stage model by their respective probabilities.
Linear + non-linear	
1. Ordinary Least Squares + Regression Trees (OLS + RT), Neural Networks (OLS + ANN)	The purpose of this two-stage technique is to combine the good comprehensibility of a linear model with the predictive power of a non-linear regression technique (Van Gestel, et al. 2005). In the first stage, a linear model is built using OLS. In the second stage, the residuals of this linear model are estimated with a non-linear regression model. This estimate for the residual is then added to the OLS estimate to obtain a more accurate prediction for LGD.

The following sections detail the formulation and considerations of the linear and non-linear techniques considered in the estimation of LGD. Depictions of the respective nodes that can be utilized in SAS Enterprise Miner are also given.

4.2.1 Ordinary Least Squares – Linear Regression

Ordinary least squares regression (Draper and Smith, 1998) is the most common technique to find optimal parameters $\mathbf{b}^T = \begin{bmatrix} b_0, b_1, b_2, \ldots, b_n \end{bmatrix}$ to fit a linear model to a data set:

$$y = \mathbf{b}^T \mathbf{x} \quad (4.1)$$

where $\mathbf{x}^T = \begin{bmatrix} 1, x_1, x_2, \ldots, x_n \end{bmatrix}$. OLS approaches this problem by minimizing the sum of squared residuals:

$$\sum_{i=1}^{l} (e_i)^2 = \sum_{i=1}^{l} \left(y_i - \mathbf{b}^T \mathbf{x}_i \right)^2 \quad (4.2)$$

By taking the derivative of this expression and subsequently setting the derivative equal to zero:

$$\sum_{i=1}^{l} \left(y_i - \mathbf{b}^T \mathbf{x}_i \right) \mathbf{x}_i^T = 0 \quad (4.3)$$

the model parameters $\mathbf{b}$ can be retrieved as:

$$\mathbf{b} = \left(\mathbf{X}^T \mathbf{X} \right)^{-1} \mathbf{X}^T \mathbf{y} \quad (4.4)$$

with $\mathbf{X}^T = \begin{bmatrix} \mathbf{x}_1, \mathbf{x}_2, \ldots, \mathbf{x}_l \end{bmatrix}$ and $\mathbf{y} = \begin{bmatrix} y_1, y_2, \ldots, y_l \end{bmatrix}^T$.

The **Regression node** (Figure 4.1) in the Model tab of Enterprise Miner can be utilized for linear regression by changing the Regression Type option on the Class Targets property panel of the node to Linear Regression. The equivalent procedure in SAS/STAT is **proc reg**.

Figure 4.1: Linear Regression Node

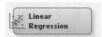

4.2.2 Ordinary Least Squares with Beta Transformation

Whereas OLS regression tests generally assume normality of the dependent variable y, the empirical distribution of LGD can often be approximated more accurately by a beta distribution (Gupton and Stein, 2002). Assuming that y is constrained to the open interval $(0,1)$, the cumulative distribution function (CDF) of a beta distribution is given by:

$$\beta(y; a, b) = \frac{\Gamma(a+b)}{\Gamma(a)\Gamma(b)} \int_0^y v^{a-1} (1-v)^{b-1} \, dv \quad (4.5)$$

where $\Gamma()$ denotes the well-known Gamma function, and a,b are two shape parameters, which can be estimated from the sample mean μ and variance σ^2 using the method of the moments:

$$a = \frac{\mu^2\left(1-\mu\right)}{\sigma^2} - \mu \; ; \; b = a\left(\frac{1}{\mu}-1\right) \;\; (4.6)$$

A potential solution to improve model fit, therefore, is to estimate an OLS model for a transformed dependent variable $y_i^* = N^{-1}\left(\beta\left(y_i;a,b\right)\right)$ $\left(i=1,...,l\right)$, in which $N^{-1}()$ denotes the inverse of the standard normal CDF. The predictions by the OLS model are then transformed back through the standard normal CDF and the inverse of the fitted beta CDF to get the actual LGD estimates.

Figure 4.2 displays the beta transformation applied in a **Transform Variables node** with an OLS model applied to the transformed target.

Figure 4.2: Combination of Beta Transformation and Linear Regression Nodes

The transformation code required to compute a beta transformation in either the **Transform Variables node** or a **SAS Code node** is detailed in Chapter 5, Section 5.4.

4.2.3 Beta Regression

Instead of performing a beta transformation prior to fitting an OLS model, an alternative beta regression approach is outlined in Smithson and Verkuilen (2006). Their preferred model for estimating a dependent variable bounded between 0 and 1 is closely related to the class of generalized linear models and allows for a dependent variable that is beta-distributed conditional on the covariates. Instead of the usual parameterization though of the beta distribution with shape parameters a,b, they propose an alternative parameterization involving a location parameter μ and a precision parameter ϕ, by letting:

$$\mu = \frac{a}{a+b} \; ; \; \phi = a+b \;\; (4.7)$$

It can be easily shown that the first parameter is indeed the mean of a $\beta\left(a,b\right)$ distributed variable, whereas

$\sigma^2 = \dfrac{\mu\left(1-\mu\right)}{\left(\phi+1\right)}$, so for fixed μ, the variance (dispersion) increases with smaller ϕ.

Two link functions mapping the unbounded input space of the linear predictor into the required value range for both parameters are then chosen via the logit link function for the location parameter (as its value must be squeezed into the open unit interval) and a log function for the precision parameter (which must be strictly positive), resulting in the following sub models:

$$\mu_i = E\left(y_i \,|\, \mathbf{x}_i\right) = \frac{e^{\mathbf{b}^T\mathbf{x}_i}}{1+e^{\mathbf{b}^T\mathbf{x}_i}} \;\; (4.8)$$

$$\phi_i = e^{-\mathbf{d}^T\mathbf{x}_i}$$

This particular parameterization offers the advantage of producing more intuitive variable coefficients (as the two rows of coefficients, $\mathbf{b}^T$ and $\mathbf{d}^T$, provide an indication of the effect on the estimate itself and its precision, respectively). By further selecting which variables to include in (or exclude from) the second sub model, one can explicitly model heteroskedasticity. The resulting log-likelihood function is then used to compute maximum-likelihood estimators for all model parameters.

The beta regression model can be formulated through the use of the SAS/STAT procedure **proc nlmixed**, and example of the code is presented below in Figure 4.3 (the example LGD_DATA needs to be standardized first using **proc standard**):

Figure 4.3: Proc nlmixed Example Code

```
proc nlmixed data =work.STNDSTANDARDIZEDLGD_DATA tech= trureg hess cov itdetails;
    title 'One-Way Layout';
/*  linear predictors;*/
    Xb = b0 + b1*stnd_INPUT1 + b2*stnd_INPUT2 + b3*stnd_INPUT3 + b4*stnd_INPUT4 + b5*stnd_INPUT5
    + b6*stnd_INPUT6 + b7*stnd_INPUT7 + b8*stnd_INPUT8
    + b9*stnd_INPUT9 + b10*stnd_INPUT10 + b11*stnd_INPUT11 + b12*stnd_INPUT12 + b13*stnd_INPUT13 ;

    Wd = d0;

/*  link functions transform linear predictors;*/
    mu = exp(Xb)/(1 + exp(Xb));
    phi = exp(-1*Wd);

/*  transform to standard parameterization for easy entry to log-likelihood;*/
    w = mu*phi;
    t = phi - mu*phi;

/*  log-likelihood;*/
    ll = lgamma(w+t) - lgamma(w) - lgamma(t) + (w-1)*log(LGD) + (t-1)*log(1-LGD);

    model stnd_LGD ~ general(ll);
    predict mu out = LGD_pred_het;
    ods output ParameterEstimates=BETA_REG_PARMS;
run;
```

The model statement of **proc nlmixed** is set to model stnd_LGD as a general likelihood function. This code can be formulated within a **SAS Code node** within the Enterprise Miner environment as shown in Figure 4.4. More information detailing the syntax for **proc nlmixed** can be found in the SAS/STAT documentation.

Figure 4.4: Beta Regression (SAS Code Node)

4.2.4 Ordinary Least Squares with Box-Cox Transformation

The aim of the family of Box-Cox transformations is to make the residuals of the regression model more homoscedastic and closer to a normal distribution (Box and Cox, 1964). The Box-Cox transformation on the dependent variable y_i takes the form

$$\begin{cases} \dfrac{\left((y_i+c)^\lambda - 1\right)}{\lambda} & \text{if } \lambda \neq 0 \\ \log(y_i+c) & \text{if } \lambda = 0 \end{cases} \quad (4.9)$$

with power parameter λ and parameter c. If needed, the value of c can be set to a non-zero value to rescale y_i so that it becomes strictly positive. After a model is built on the transformed dependent variable using OLS, the predicted values can be transformed back to their original value range.

An example code for the application of a Box-Cox transformation using **proc transreg** is shown in Program 4.1, using the provided LGD_DATA data set:

Program 4.1: Box-Cox Transreg Code

```
%let inputs = Input1-Input13;

proc transreg data=output.LGD_DATA ss2 details;
     model BoxCox(LGD / lambda=-3 to 3 by 0.25) =
            identity(&inputs);
   run;
```

For the above code, the value of c is set to 0 with lambda varied from -3 to 3 by 0.25. As with the OLS, with beta transformation approach (see Section 4.2.2) a Box-Cox transformation can be applied in either a SAS Code node or Transform Variables node prior to an OLS model applied to the transformed target. Figure 4.5 displays the Box-Cox transformation applied in a Transform Variables node with an OLS model applied to the transformed target.

Figure 4.5: Combination of Box-Cox Transformation and Linear Regression Nodes

4.2.5 Regression Trees

In Chapter 3, we looked at the application of decision trees for classification problems. Decision trees can also be used for regression analysis where they are designed to approximate real-valued functions as opposed to a classification task. A commonly applied impurity measure $i(t)$ for regression trees is the mean squared error or variance for the subset of observations falling into node t. Alternatively, a split may be chosen based on the p-value of an ANOVA F-test comparing between-sample variances against within-sample variances for the subsamples associated with its respective child nodes (ProbF criterion).

Regression trees (Figure 4.6) have a useful application in the prediction of the un-winsorised LGD interval distribution, as they are more capable of dealing with the extreme values that are often present.

Figure 4.6: Decision Trees Node (Renamed Regression Trees)

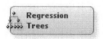

Late in this chapter, we explore a comparison of Regression Trees against traditional uses of linear regression models.

4.2.6 Artificial Neural Networks

Figure 4.7: Neural Networks Node (Relabeled Artificial Neural Network)

In Chapter 3, we also detailed the implementation of Neural Networks for classification problems. In terms of regression, Neural Networks (Figure 4.7) produce an output value by feeding inputs through a network whose subsequent nodes apply some chosen activation function to a weighted sum of incoming values. The type of ANN assessed in this chapter is the popular multilayer perceptron (MLP).

4.2.7 Linear Regression and Non-linear Regression

Techniques such as Neural Networks are often seen as "black box" techniques, meaning that the model obtained is not understandable in terms of physical parameters. This is an obvious issue when applying these techniques to a credit risk modeling scenario where physical parameters are required. To solve this issue, a two-stage approach to combine the good comprehensibility of OLS with the predictive power of a non-linear regression technique can be used (Van Gestel, et al. 2005). In the first stage, an ordinary least squares regression model is built:

$$y = \mathbf{b}^T\mathbf{x} + e \quad (4.10)$$

In the second stage, the residuals e of this linear model:

$$e = f(\mathbf{x}) + e^* \quad (4.11)$$

are estimated with a non-linear regression model $f(\mathbf{x})$ in order to further improve the predictive ability of the model. Doing so, the model takes the following form:

$$y = \mathbf{b}^T\mathbf{x} + f(\mathbf{x}) + e^* \quad (4.12)$$

Where e^* are the new residuals of estimating e. This chapter looks at combination of OLS with RT (Figure 4.8) and ANN.

Figure 4.8: Linear Regression and Regression Trees Nodes

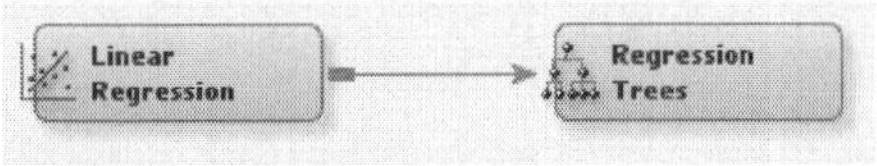

4.2.8 Logistic Regression and Non-linear Regression

The LGD distribution is often characterized by a large peak around $LGD = 0$. This non-normal distribution can lead to inaccurate regression models. This proposed two-stage technique attempts to resolve this issue by modeling the peak separately from the rest. Therefore, the first stage of this two-stage model consists of a logistic regression to estimate whether $LGD \leq 0$ or $LGD > 0$.

In a second stage, the mean of the observed values of the peak is used as prediction in the first case and a one-stage (non)linear regression model is used to provide a prediction in the second case. The latter is trained on part of the data set, which are those observations that have a $LGD > 0$. More specifically, a logistic regression results in an estimate of the probability P of being in the peak:

$$P = \frac{1}{1 + e^{-(\mathbf{b}^T\mathbf{x})}} \quad (4.13)$$

with $(1-P)$ as the probability of not being in the peak. An estimate for LGD is then obtained by:

$$y = P.\overline{y}_{peak} + (1-P).f(\mathbf{x}) + e \quad (4.14)$$

where $\overline{y}_{peak}$ is the mean of the values of $y \leq 0$, which is equitable to 0, and $f(\mathbf{x})$ is a one-stage (non)linear regression model, built on those observations only that are not in the peak. A combination of logistic regression with all aforementioned one-stage techniques (Figure 4.9) as described above, is assessed is this chapter (Matuszyk et al. 2010).

Figure 4.9: Logistic Regression and Linear Regression Nodes

4.3 Performance Metrics for LGD

Performance metrics evaluate to which degree the predictions $f(\mathbf{x}_i)$ differ from the observations y_i of the dependent variable LGD. Each of the following metrics, listed in Table 4.2, has its own method to express the predictive performance of a model as a quantitative value. The second and third columns of the table show the metric values for, respectively, the worst and best possible prediction performance. The final column shows whether the metric measures calibration or discrimination (Van Gestel and Baesens, 2009). Calibration indicates how close the predictive values are with the observed values, whereas discrimination refers to the ability to provide an ordinal ranking of the dependent variable considered. A good ranking does not necessarily imply a good calibration.

Table 4.2: Regression Performance Metrics

Metric	Worst	Best	Measure
RMSE	$+\infty$	0	Calibration
MAE	$+\infty$	0	Calibration
AUC	0.5	1	Discrimination
AOC	$+\infty$	0	Calibration
R^2	0	1	Calibration
r	0	1	Discrimination
ρ	0	1	Discrimination
τ	0	1	Discrimination

Note that the R^2 measure defined here could possibly lie outside the [0, 1] interval when applied to non-OLS models. Although alternative generalized goodness-of-fit measures have been put forward for evaluating various non-linear models, the measure defined in Table 4.2 has the advantage that it is widely used and can be calculated for all techniques (Nagelkerke, 1991).

Commonly used performance metrics for evaluating the predictive power of regression techniques include RMSE (Draper and Smith, 1998), Mean Absolute Error (Draper and Smith, 1998), Area Under the Receiver Operating Curve (AUC) (Fawcett, 2006), and correlation metrics, Pearson's Correlation Coefficient (r), Spearman's Correlation Coefficient (ρ) and Kendall's Correlation Coefficient (τ) (Cohen, et al. 2002). The most often reported performance metric is the R-square (R^2) (Draper and Smith, 1998).

4.3.1 Root Mean Squared Error

RMSE (see for example, Draper and Smith, 1998) is defined as the square root of the average of the squared difference between predictions and observations:

$$RMSE = \sqrt{\frac{1}{l}\sum_{i=1}^{l}\left(f(\mathbf{x}_i) - y_i\right)^2} \quad (4.15)$$

RMSE has the same units as the independent variable being predicted. Since residuals are squared, this metric heavily weights outliers. The smaller the value of RMSE, the better the prediction, with 0 being a perfect prediction. Here, the number of observations is given by l.

4.3.2 Mean Absolute Error

MAE (see for example, Draper and Smith, 1998) is given by the averaged absolute differences of predicted and observed values:

$$MAE = \frac{1}{l}\sum_{i=1}^{l}\left|f\left(\mathbf{x}_i\right)-y_i\right| \quad (4.16)$$

Just like RMSE, MAE has the same unit scale as the dependent variable being predicted. Unlike RMSE, MAE is not that sensitive to outliers. The metric is bound between the maximum absolute error and 0 (perfect prediction).

4.3.3 Area Under the Receiver Operating Curve

ROC curves are normally used for the assessment of binary classification techniques (see for example, Fawcett, 2006). It is, however, used in this context to measure how good the regression technique is in distinguishing high values from low values of the dependent variable. To build the ROC curve, the observed values are first classified into high and low classes using the mean $\overline{y}$ of the training set as reference. An example of an ROC curve is depicted in Figure 4.10:

Figure 4.10: Example ROC Curve

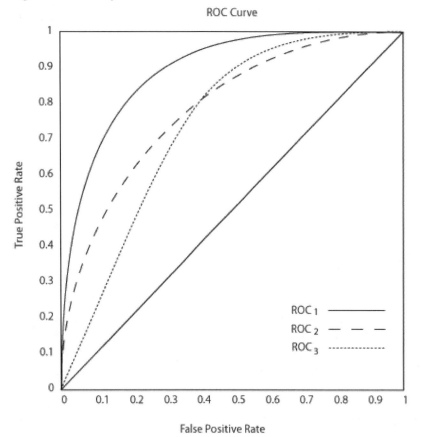

The ROC chart in Figure 4.10 graphically displays the True Positive Rate (Sensitivity) versus the False Positive Rate (1-Specificity). The true positive rate and the false positive rate are both measures that depend on the selected cutoff value of the posterior probability. Therefore, the ROC curve is calculated for all possible cutoff values. The diagonal line represents the trade-off between the sensitivity and (1-specificity) for a random model, and has an AUC of 0.5. For a well performing classifier, the ROC curve needs to be as far to the top left-hand corner as possible.

If we consider the example curves ROC_1, ROC_2 and ROC_3, each point on the curves represents cutoff probabilities. Points that are closer to the upper right corner correspond to low cutoff probabilities. Points that are closer to the lower left corner correspond to high cutoff probabilities. The extreme points (1,1) and (0,0) represent rules where all cases are classified into either class 1 (event) or class 0 (non-event). For a given false positive rate (the probability of a non-event that was predicted as an event), the curve indicates the corresponding true positive rate, the probability for an event to be correctly predicted as an event. Therefore, for a given false positive rate on the False Positive Rate axis, the true positive rate should be as high as possible. The different curves in the chart exhibit various degrees of concavity. The higher the degree of concavity, the better the model is expected to be. In Figure 4.10, ROC_1 appears to be the best model. Conversely, a poor model of random predictions appears as a flat 45-degree line. Curves that push upward and to the left represent better models.

In order to compare the ROC curves of different classifiers, the area under the receiver operating characteristic curve (AUC) must be computed. The AUC statistic is similar to the Gini Coefficient which is equal to $Gini = 2AUC - 1$.

4.3.4 Area Over the Regression Error Characteristic Curves

We also evaluate the Area Over the Regression Error Characteristic (REC) curves performance metric in this chapter. This statistic is often simplified to Area Over the Curve (AOC). Figure 4.11 displays an example of three techniques plotted against a mean line:

Figure 4.11: Example REC Curve

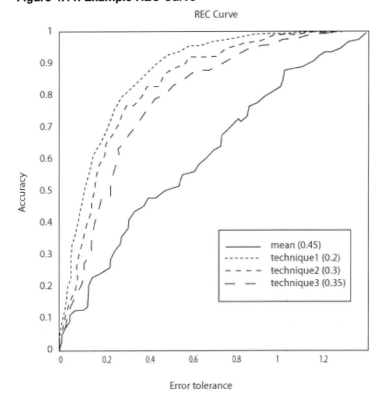

REC curves (Bi and Bennet, 2003) generalize ROC curves for regression. The AOC curve plots the error tolerance on the x-axis versus the percentage of points predicted within the tolerance (or accuracy) on the y-axis (Figure 4.11). The resulting curve estimates the cumulative distribution function of the squared error. The area over the REC curve (AOC) is an estimate of the predictive power of the technique. The metric is bound between 0 (perfect prediction) and the maximum squared error.

4.3.5 R-square

R-square R^2 (see for example, Draper and Smith, 1998) can be defined as 1 minus the fraction of the residual sum of squares to the total sum of squares:

$$R^2 = 1 - \frac{SS_{err}}{SS_{tot}} \quad (4.17)$$

where $SS_{err} = \sum_{i=1}^{l}\left(y_i - f\left(\mathbf{x}_i\right)\right)^2$, $SS_{tot} = \sum_{i=1}^{l}\left(y_i - \overline{y}\right)^2$ and $\overline{y}$ is the mean of the observed values. Since

the second term in the formula can be seen as the fraction of unexplained variance, the R^2 can be interpreted as the fraction of explained variance. Although R^2 is usually expressed as a number on a scale from 0 to 1, R^2 can yield negative values when the model predictions are worse than using the mean $\overline{y}$ from the training set as prediction. Although alternative generalized goodness-of-fit measures have been put forward for evaluating various non-linear models (see Nagelkerke, 1991), R^2 has the advantage that it is widely used and can be calculated for all techniques.

4.3.6 Pearson's Correlation Coefficient

Pearson's r (see Cohen, et al. 2002) is defined as the sum of the products of the standard scores of the observed and predicted values divided by the degrees of freedom:

$$r = \frac{1}{l-1}\sum_{i=1}^{l}\left(\frac{y_i - \overline{y}}{s_y}\right)\left(\frac{f\left(\mathbf{x}_i\right) - \overline{x}}{s_f}\right) \quad (4.18)$$

with $\overline{y}$ and $\overline{x}$ the mean and s_y and s_f the standard deviation of respectively the observations and predictions. Pearson's r can take values between -1 (perfect negative correlation) and +1 (perfect positive correlation) with 0 meaning no correlation at all.

4.3.7 Spearman's Correlation Coefficient

Spearman's ρ (see Cohen, et al. 2002) is defined as Pearson's r applied to the rankings of predicted and observed values. If there are no (or very few) tied ranks, however, it is common to use the equivalent formula:

$$\rho = 1 - \frac{6\sum_{i=1}^{l} d_i^2}{l\left(l^2 - 1\right)} \quad (4.19)$$

where d_i is the difference between the ranks of observed and predicted values. Spearman's ρ can take values between -1 (perfect negative correlation) and +1 (perfect positive correlation) with 0 meaning no correlation at all.

4.3.8 Kendall's Correlation Coefficient

Kendall's τ (see Cohen, et al. 2002) measures the degree of correspondence between observed and predicted values. In other words, it measures the association of cross tabulations:

$$\tau = \frac{n_c - n_d}{\frac{1}{2}l(l-1)} \quad (4.20)$$

where n_c is the number of concordant pairs and n_d is the number of discordant pairs. A pair of observations $\{i,k\}$ is said to be concordant when there is no tie in either observed or predicted LGD ($y_i \neq y_k$, $f(\mathbf{x}_i) \neq f(\mathbf{x}_k)$), and if $\text{sign}(f(\mathbf{x}_k) - f(\mathbf{x}_i)) = \text{sign}(y_k - y_i)$, where $i,k = 1,\ldots,l\,(i \neq k)$.

Similarly, it is said to be discordant if there is no tie and if $\text{sign}(f(\mathbf{x}_k) - f(\mathbf{x}_i)) = -\text{sign}(y_k - y_i)$.

Kendall's τ can take values between -1 (perfect negative correlation) and +1 (perfect positive correlation) with 0 meaning no correlation is present.

4.4 Model Development

In this section, we explore the development process of an LGD model utilizing SAS Enterprise Miner and SAS/STAT techniques. The characteristics of the data sets used in an experimental model build framework to assess the predictive performance of the regression techniques are also given. Further, a description of a technique's parameter setting and tuning is provided where required.

4.4.1 Motivation for LGD models

As discussed at the start of this chapter, the purpose of developing LGD models is to categorize the customers into risk groups based on their historical information. Through a careful monitoring and assessment of the customer accounts, the risk related characteristics can be determined. As detailed in Chapter 1, banks require an internal LGD estimate under the Advanced IRB approach to calculate their risk-weighted assets (RWA) based on these risk characteristics. In addition to acting as a function of RWA, LGD predictions can be used as input variables to estimate potential credit losses. The LGD values assigned to each customer are used for early detection of high-risk accounts and to enable organizations to undertake targeted interventions with the at risk customers. By using analytics for detecting these high-risk accounts organizations can reduce their overall portfolio risk levels and enables a management of the credit portfolios more effectively.

4.4.2 Developing an LGD Model

4.4.2.1 Overview

SAS can be utilized in the creation of LGD models through the application of statistical techniques either within SAS/STAT or SAS Enterprise Miner. As with PD estimation, discussed in Chapter 3, before an analyst creates an LGD model, they must determine the time frame and the analytical base table variables. The data must be first prepared in order to deal with duplicate, missing, or outlier values as discussed in Chapter 2, and to select a suitable time frame:

Figure 4.12: Outcome Time Frame for an LGD Model

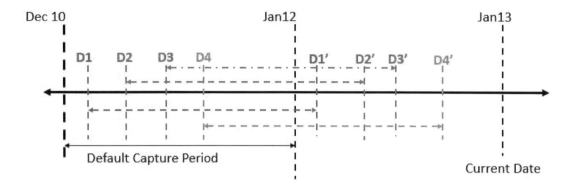

The above Figure 4.12 details a typical outcome time frame for determining qualifying accounts for use within the LGD modeling process. The default capture period is the period during which all accounts or credit facilities that have defaulted are considered for analysis. In the example time frame, D1-D4 are the default dates of accounts 1-4 respectively, and the time between D1 and D1' is the recovery period for account 1.

In order to determine the value of LGD, all customer accounts that have observed a default during the capture period have their recovery cash flows summed in the recovery cycle. The recovery cycle is the period of time that starts from the data of default for each account and ends after a defined period. When a default occurs, the actual loss incurred by the accounts is determined. The recovery amount is determined as the acquisition of the sum due to the financial institution by the customer. The recovery amount is, therefore, a proportion of the initial loaned amount at time of application.

4.4.2.2 Model Creation Process Flow

The following Figure 4.13 shows the model creation process flow for an example LGD model that is provided with this book. The subsequent sections describe nodes in the process flow diagram and the step-by-step approach an analyst can take in the development of an LGD model.

Figure 4.13: LGD Model Flow

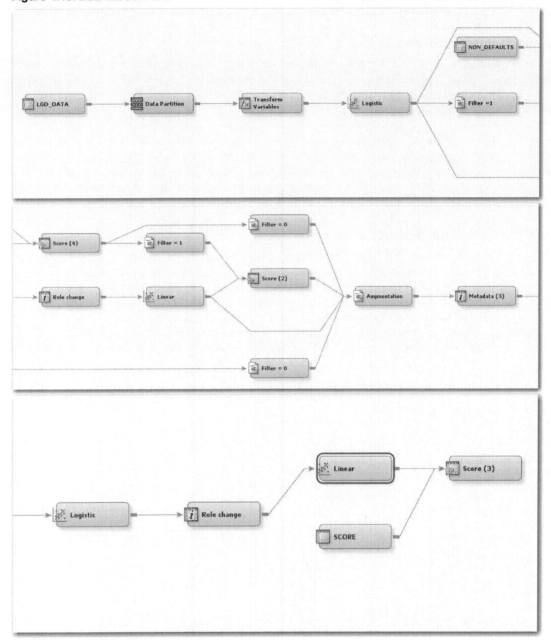

4.4.2.3 LGD Data

A typical LGD development data set contains independent variables such as LGD, EAD, recovery cost, value at recovery, and default date. In the model flow in SAS Enterprise Miner, the indicator variable is created in the **SAS Code node** (Filter = 1 and Filter = 0) for separating the default and non-default records. For example, if the default date for an account is missing, then that account is considered non-default. If a default date for an account is available, then that account is considered default. If the account has caused a loss to the bank, then the target variable has a value of 1; otherwise, it has a value of 0. A **Data Partition node** is also utilized to split the initial LGD_DATA sample into a 70% train set and a 30% validation set.

4.4.2.4 Logistic Regression Model

The application of a **Logistic Regression node** attempts to predict the probability that a binary or ordinal target variable will attain the event of interest based on one or more independent inputs. By using LGD as a binary target variable, logistic regression is applied on the defaulted records. A binary flag for LGD is created in the **Transform Variables node** after the data partition, where LGD≤0 is equal to 0 and LGD>0 is equal to 1.

4.4.2.5 Scoring Non-Defaults

The development LGD_DATA sample contains only accounts that have experienced a default along with an indication as to where recoveries have been made. From this information, a further LOSS_FLG binary target variable can be calculated, where observations containing a loss amount at the point of default receive a 1 and those that do not contain a loss amount equal to 0.

At this point, we are not considering the non-defaulting accounts, as a target variable does not exist for these customers. As with rejected customers (see Chapter 3), not considering non-defaulting accounts would bias the results of the final outcome model, as recoveries for non-defaulting accounts may not yet be complete and could still incur loss. To adjust for this bias, the non-defaulting account data can be scored with the model built on the defaulted records to achieve an inferred binary LOSS_FLG of 0 or 1.

4.4.2.6 Predicting the Amount of Loss

Once the regression model has been built to predict the LOSS_FLG for the defaulted accounts (4.4.2.4) and an inferred LOSS_FLG has been determined for the non-defaulting accounts (4.4.2.5), the next stage of the LGD model development is to predict the actual value of loss. As the amount of loss is an interval amount, traditionally, linear regression is applied in the prediction of the loss value. The difficulty that exists with the prediction of the LGD value is the distribution LGD presents. Typically, and through the understanding of historical LGD distributions (see Figure 4.15 for further examples of LGD distributions) shows that LGD tends to have a bimodal distribution which often exhibits a peak near 0 and a smaller peak around 1 as show in the following Figure 4.14. (A discussion of the techniques and transformations that can be utilized to mitigate for this distribution are presented in Section 4.2 and applied in Section 4.5).

Figure 4.14: Example LGD Distribution

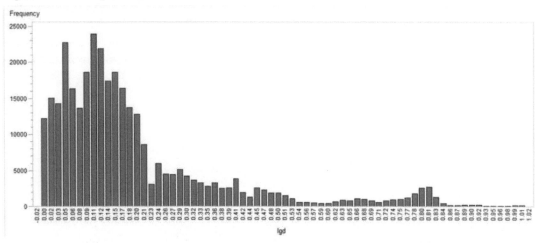

In the LGD Model Flow presented in Figure 4.13, a linear regression model using the **Regression node** in Enterprise Miner is applied to those accounts that have a LOSS_FLG of 1. As with Section 4.4.2.5, this model can also be applied to the non-defaulting accounts to infer the expected loss amount for these accounts. A final augmented data set can be formed by appending the payers/non-payers for the defaulted accounts and the inferred payers/non-payers for the non-defaulting accounts.

The augmented data set is then remodeled using the LOSS_FLAG as the binary dependent target with a logistic regression model. The total number of payers from the augmented data are filtered. Another linear regression model is applied on the filtered data to predict the amount of loss. While scoring, only the non-defaulted accounts are scored.

4.4.2.7 Model Validation

A number of model validation statistics are automatically calculated within the model nodes and **Model Comparison node** within SAS Enterprise Miner. A common evaluation technique for LGD is the Area Under the Receiver Operating Characteristic Curve, or R-square value. By using a validation sample within the Miner project, performance metrics will be computed across both the training and validation samples. To access these validation metrics in the model flow diagram, right-click a modeling node and select Results. By default for the **Regression node,** a Score Rankings Overlay plot and Effects Plot are displayed. Additional plots such as the Estimate Selection Plot, useful for identifying at which step an input variable entered the model using Stepwise, can be accessed by clicking View ▶ Model or View ▶ Assessment in the Results screen of the node. In order to validate a model, it is important to examine how well the trained model fits to the unseen validation data. In the Score Rankings plot, any divergence between the Train line and Validate line identifies where the model is unable to generalize to new data. Chapter 7 describes a more comprehensive list of LGD model validation metrics along with code to calculate additional performance graphics.

4.5 Case Study: Benchmarking Regression Algorithms for LGD

In this section, an empirical case study is given to demonstrate how well the regression algorithms discussed perform in the context of estimating LGD. This study comprises of the author's contribution to a larger study, which can be found in Loterman et al. (2009).

4.5.1 Data Set Characteristics

Table 4.3 displays the characteristics of 6 real life lending LGD data sets from a series of financial institutions, each of which contains loan-level data about defaulted loans and their resulting losses. The number of data set entries varies from a few thousands to just under 120,000 observations. The number of available input variables ranges from 12 to 44. The types of loan data set included are personal loans, corporate loans, revolving credit, and mortgage loans. The empirical distribution of LGD values observed in each of the data sets is displayed in Figure 4.15. Note that the LGD distribution in consumer lending often contains one or two spikes around $LGD = 0$ (in which case there was a full recovery) and/or $LGD = 1$ (no recovery). Also, a number of data sets include some LGD values that are negative (because of penalties paid, gains in collateral sales, etc.) or larger than 1 (due to additional collection costs incurred); in other data sets, values outside the unit interval were truncated to 0 or 1 by the banks themselves. Importantly, LGD does not display a normal distribution in any of these data sets.

Table 4.3: Data Set Characteristics of Real Life LGD Data

Data set	Type	Inputs	Data Set Size	Training Set Size	Test Set Size
BANK 1	Personal loans	44	47,853	31,905	15,948
BANK 2	Mortgage loans	18	119,211	79,479	39,732
BANK 3	Mortgage loans	14	3,351	2,232	1,119
BANK 4	Revolving credit	12	7,889	5,260	2,629
BANK 5	Mortgage loans	35	4,097	2,733	1,364
BANK 6	Corporate loans	21	4,276	2,851	1,425

Figure 4.15: LGD Distributions of Real Life LGD Data Sets

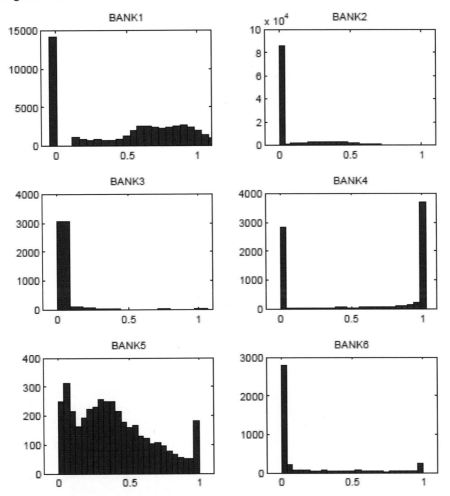

4.5.2 Experimental Set-Up

First, each data set is randomly shuffled and divided into two-thirds training set and one-third test set. The training set is used to build the models while the test set is used solely to assess the predictive performance of these models. Where required, continuous independent variables are standardized with the sample mean and standard deviation of the training set. Nominal and ordinal independent variables are encoded with dummy variables.

An input selection method is used to remove irrelevant and redundant variables from the data set, with the aim of improving the performance of regression techniques. For this, a stepwise selection method is applied for building the linear models. For computational efficiency reasons, an R^2 based filter method (Freund and Littell, 2000) is applied prior to building the non-linear models.

After building the models, the predictive performance of each data set is measured on the test set by comparing the predictions and observations according to several performance metrics. Next, an average ranking of techniques over all data sets is generated per performance metric as well as a meta-ranking of techniques over all data sets and all performance metrics.

Finally, the regression techniques are statistically compared with each other (Demšar, 2006). A Friedman test is performed to test the null hypothesis that all regression techniques perform alike according to a specific performance metric, i.e., performance differences would just be due to random chance (Friedman, 1940). A more detailed summary and the applied formulas can be found in the previous chapter (Section 4.3.4).

4.5.2.1 Parameter Settings and Tuning

During model building, several techniques require parameters to be set or tuned. This section describes how these are set or tuned where appropriate. No additional parameter tuning was required for the Linear Regression (OLS), Linear Regression with Beta Transformation (B-OLS), and Beta Regression (BR) techniques.

4.5.2.2 Ordinary Least Squares with Box-Cox Transformation (BC-OLS)

The value of parameter c is set to zero. The value of the power parameter λ is varied over a chosen range (from -3 to 3 in 0.25 increments) and an optimal value is chosen based on a maximum likelihood criterion.

4.5.2.3 Regression Trees (RT)

For the regression tree model, the training set is further split into a training and validation set. The validation set is used to select the criterion for evaluating candidate splitting rules (variance reduction or ProbF), the depth of the tree, and the threshold p-value for the ProbF criterion. The choice of tree depth, the threshold p-value for the ProbF criterion and criterion method was selected based on the mean squared error on the validation set.

4.5.2.4 Artificial Neural Networks (ANN)

For the artificial neural networks (ANN) model, the training set is further split into a training and validation set. The validation is used to evaluate the target layer activation functions (logistic, linear, exponential, reciprocal, square, sine, cosine, tanh and arcTan) and number of hidden neurons (1-20) used in the model. The weights of the network are first randomly initialized and then iteratively adjusted so as to minimize the mean squared error. The choice of activation function and number of hidden neurons is selected based on the mean squared error on the validation set. The hidden layer activation function is set as logistic.

4.5.3 Results and Discussion

Table 4.4 shows the performance results obtained for all techniques on the BANK 2 data for illustrative purposes. The best performing model according to each metric is underlined. Figure 4.16 displays a series of box plots for the observed distributions of performance values for the metrics AUC, R^2, r, ρ and τ. Similar trends can be observed across all metrics. Note that differences in type of data set, number of observations, and available independent variables are the likely causes of the observed variability of actual performance levels between the 6 different data sets.

Although all of the performance metrics listed above are useful measures in their own right, it is common to use the R-square R^2 to compare model performance, since R^2 measures calibration and can be compared meaningfully across different data sets. It was found that the average R^2 of the models varies from about 4% to 43%. In other words, the variance in LGD that can be explained by the independent variables is consistently below 50%, implying that most of the variance cannot be explained even with the best models. Note that although R^2 usually is a number on a scale from 0 to 1, R^2 can yield negative values for non-OLS models when the model predictions are worse than always using the mean from the training set as prediction.

Table 4.4: BANK 2 Performance Results

Technique	MAE	RMSE	AUC	AOC	R^2	r	ρ	τ
OLS	0.1187	0.1613	0.8100	0.0259	0.2353	0.4851	0.4890	0.3823
B-OLS	0.1058	0.1621	0.8000	0.0262	0.2273	0.4768	0.4967	0.3881
BC-OLS	0.1056	0.1623	0.7450	0.0262	0.2226	0.4718	0.4990	0.3900
BR	0.1020	0.1661	0.7300	0.0275	0.2120	0.4635	0.4857	0.3861
RT	0.0978	0.1499	0.7710	0.0224	0.3390	0.5823	0.5452	0.4357
ANN	<u>0.0956</u>	<u>0.1472</u>	0.8530	<u>0.0216</u>	<u>0.3632</u>	<u>0.6029</u>	0.5549	0.4366
LOG+OLS	0.1060	0.1622	0.7590	0.0255	0.2268	0.4838	0.5206	0.4084

Technique	MAE	RMSE	AUC	AOC	R^2	r	ρ	τ
LOG+B-OLS	0.1040	0.1567	0.8320	0.0245	0.2779	0.5286	0.5202	0.4083
LOG+BC-OLS	0.1034	0.1655	0.7320	0.0273	0.2124	0.4628	0.4870	0.3820
LOG+BR	0.1015	0.1688	0.7250	0.0285	0.2024	0.4529	0.4732	0.3876
LOG+RT	0.1041	0.1538	0.8360	0.0236	0.3049	0.5545	0.5254	0.4126
LOG+ANN	0.1011	0.1531	0.8430	0.0234	0.3109	0.5585	0.5380	0.4240
OLS+RT	0.1015	0.1506	0.8410	0.0227	0.3331	0.5786	0.5344	0.4188
OLS+ANN	0.0999	0.1474	<u>0.8560</u>	0.0217	0.3612	0.6010	<u>0.5585</u>	<u>0.4398</u>

Figure 4.16: Comparison of Predictive Performances Across Six Real Life Retail Lending Data Sets

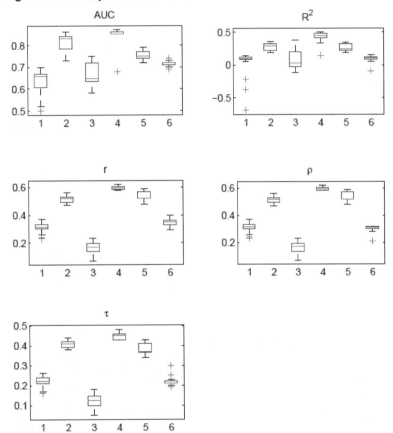

The linear models that incorporate some form of transformation to the dependent variable (B-OLS, BR, BC-OLS) are shown to perform consistently worse than OLS, despite the fact that these approaches are specifically designed to cope with the violation of the OLS normality assumption. This suggests that they too have difficulties dealing with the pronounced point densities observed in LGD data sets, while they may be less efficient than OLS or they could introduce model bias if a transformation is performed prior to OLS estimation (as is the case for B-OLS and BC-OLS).

Perhaps the most striking result is that, in contrast with prior benchmarking studies on classification models for PD (Baesens, et al. 2003), non-linear models such as ANN significantly outperform most linear models in the prediction of LGD. This implies that the relation between LGD and the independent variables in the data sets is non-linear. Also, ANN generally performs better than RT. However, ANN results in a black-box model while RT has the ability to produce comprehensible white-box models. To circumvent this disadvantage, one could try to obtain an interpretation for a well-performing black-box model by applying rule extraction techniques (Martens, et al. 2007, Martens, et al. 2009).

The performance evaluation of the class of two-stage models in which a logistic regression model is combined with a second-stage (linear or non-linear) model (LOG +), is less straightforward. Although a weak trend is noticeable that logistic regression combined with a linear model tends to increase the performance of the latter, it appears that logistic regression combined with a non-linear model slightly reduces the strong performance of the latter. Because the LGD distributions from BANK4, BANK5, and BANK6 also show a peak at $LGD = 1$, the performance of these models could possibly be increased by slightly altering the technique. Replacing the binary logistic regression component by an ordinal logistic regression model distinguishing between 3 classes ($LGD \leq 0, 0 < LGD < 1, LGD \geq 1$) and then using a second-stage model for $0 < LGD < 1$ could perhaps better account for the presence of both peaks.

In contrast with the previous two-stage model, a clear trend can be observed for the combination of a linear and a non-linear model (OLS +). By estimating the error residual of an OLS model using a non-linear technique, the prediction performance tends to increase to somewhere very near the level of the corresponding one-stage non-linear technique. What makes these two-stage models attractive is that they have the advantage of combing the high prediction performance of non-linear regression with the comprehensibility of a linear regression component. Note that this modeling method has also been successfully applied in a PD modeling context (Van Gestel, et al. 2005, Van Gestel, et al. 2006, Van Gestel, et al. 2007).

The average ranking across all data sets according to each performance metric is listed in Table 4.5. The best performing technique for each metric is underlined and techniques that perform significantly worse than the best performing technique for that metric according to the Nemenyi's post-hoc test ($\alpha = 0.5$) are italicized. The last column illustrates the meta-ranking (MR) as the average ranking (AR) across all data sets and across all metrics. The techniques in the table are sorted according to their meta-ranking. Additionally, columns MR_{cal} and MR_{dis} cover the meta-ranking only including calibration and discrimination metrics, respectively. The best performing techniques are consistently ranked in the top according to each metric, no matter whether they measure calibration or discrimination.

The results of the Friedman test and subsequent Nemenyi's post-hoc test with significance level ($\alpha = 0.5$) can be intuitively visualized using Demšar's significance diagram (Demšar, 2006). Figure 4.17 display the Demšar significance diagrams for the AOC and R^2 metric ranks across all 6 data sets. The diagrams display the performance rank of each technique along with a line segment representing its corresponding critical difference (CD = 10.08).

Table 4.5: Average Rankings (AR) and Meta-Rankings (MR) Across All Metrics and Data Sets

Rank	Technique	MAE	RMSE	AUC	AOC	R^2	r	ρ	τ	MR_{cal}	MR_{dis}	MR
1	ANN	3.2	2.8	5.0	2.5	2.7	3.1	7.0	7.1	2.8	5.5	4.2
2	LOG+ANN	6.0	4.2	6.8	4.1	4.2	4.2	6.3	6.5	4.6	6.0	5.3
3	OLS+ANN	9.0	6.5	3.6	4.7	4.3	4.3	6.2	6.3	6.1	5.1	5.6
4	OLS+RT	6.8	4.3	5.3	6.0	6.0	6.2	7.0	7.7	5.8	6.5	6.2
5	RT	8.6	7.0	12.9	7.4	7.0	7.8	7.3	4.7	7.5	8.2	7.8
6	LOG+RT	9.7	9.4	10.0	9.4	9.3	9.3	9.3	9.2	9.5	9.5	9.5
7	LOG+OLS	12.6	10.3	10.7	9.8	9.9	11.7	11.7	11.8	10.6	11.5	11.1
8	LOG+B-OLS	6.5	12.0	11.8	12.0	12.0	11.2	13.1	13.3	10.6	12.3	11.5
9	OLS	*13.9*	10.6	9.3	10.5	10.5	11.8	12.8	13.3	11.4	11.8	11.6
10	B-OLS	7.8	*14.3*	10.3	*14.8*	*14.0*	*13.8*	11.0	11.3	12.7	11.6	12.2
11	LOG+BC-OLS	8.2	*14.1*	13.0	*14.1*	*13.7*	13.0	11.8	11.8	12.5	12.4	12.5
12	BC-OLS	9.8	*15.7*	13.8	*15.6*	*15.3*	*14.7*	10.7	11.0	14.1	12.5	*13.3*
13	BR	*14.5*	12.9	*14.1*	13.3	*15.3*	*14.5*	11.8	11.5	14.0	13.0	*13.5*
14	LOG+BR	*13.9*	*14.9*	*14.9*	15.0	*14.6*	*14.2*	*14.2*	13.8	14.6	14.3	*14.4*

Despite clear and consistent differences between regression techniques in terms of R^2, most techniques do not differ significantly according to the Nemenyi test. Nonetheless, failing to reject the null hypothesis that two techniques have equal performances does not guarantee that it is true. For example, Nemenyi's test is unable to reject the null hypothesis that ANN and OLS have equal performances, although ANN consistently performs better than OLS. This can mean that the performance differences between these two are just due to chance, but the result could also be a Type II error. Possibly, the Nemenyi test does not have sufficient power to detect a significant difference, given a significance level of ($\alpha = 0.5$), 6 data sets, and 14 techniques. The insufficient power of the test can be explained by the use of a large number of techniques in contrast with a relatively small number of data sets.

Figure 4.17: Demšar's Significance Diagram for AOC and R^2 Based Ranks Across Six Data Sets

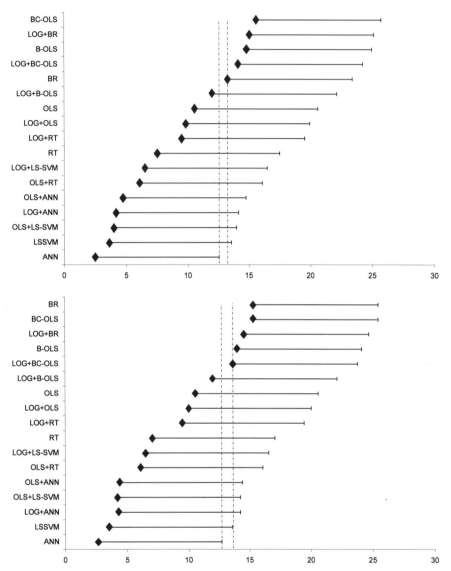

4.6 Chapter Summary

In this chapter, the processes and best-practices for the development of a Loss Given Default model using SAS Enterprise Miner and SAS/STAT have been given.

A full development of comprehensible and robust regression models for the estimation of Loss Given Default (LGD) for consumer credit has been detailed. An in-depth analysis of the predictive variables used in the modeling of LGD has also been given, showing that previously acknowledged variables are significant and identifying a series of additional variables.

This chapter also evaluated a case study into the estimation of LGD through the use of 14 regression techniques on six real life retail lending data sets from major international banking institutions. The average predictive performance of the models in terms of R^2 ranges from 4% to 43%, which indicates that most resulting models do not have satisfactory explanatory power. Nonetheless, a clear trend can be seen that non-linear techniques such as artificial neural networks in particular give higher performances than more traditional linear techniques. This indicates the presence of non-linear interactions between the independent variables and the LGD, contrary to some studies in PD modeling where the difference between linear and non-linear techniques is not that explicit (Baesens, et al. 2003). Given the fact that LGD has a bigger impact on the minimal capital requirements than PD, we demonstrated the potential and importance of applying non-linear techniques for LGD modeling, preferably in a two-stage context to obtain comprehensibility as well. The findings presented in this chapter also go some way in agreeing with the findings presented in Qi and Zhao, where it was shown that non-parametric techniques such as regression trees and neural networks gave improved model fit and predictive accuracy over parametric methods (2011).

From experience in recent history, a large European bank has gone through an implementation of a two-stage modeling methodology using a non-linear model, which was subsequently approved by their respective financial governing body. As demonstrated in the above case study, if a 1% improvement in the estimation of LGD was realized, this could equate to a reduction in RWA in the region of £100 million and EL of £7 million for large retail lenders. Any reduction in RWA inevitably means more money, which is then available to lend to customers.

4.7 References and Further Reading

Acharya, V., and Johnson, T. 2007. "Insider trading in credit derivatives." Journal of Financial Economics, 84, 110–141.

Altman, E. 2006. "Default Recovery Rates and LGD in Credit Risk Modeling and Practice: An Updated Review of the Literature and Empirical Evidence." http://people.stern.nyu.edu/ealtman/UpdatedReviewofLiterature.pdf.

Baesens, B., Van Gestel, T., Viaene, S., Stepanova, M., Suykens, J. and Vanthienen, J. 2003. "Benchmarking state-of-the-art classification algorithms for credit scoring." Journal of the Operational Research Society, 54(6), 627-635.

Basel Committee on Banking Supervision. 2005. "Basel committee newsletter no. 6: validation of low-default portfolios in the Basel II framework." Technical Report, Bank for International Settlements.

Bastos, J. 2010. "Forecasting bank loans for loss-given-default. Journal of Banking & Finance." 34(10), 2510-2517.

Bellotti, T. and Crook, J. 2007. "Modelling and predicting loss given default for credit cards." Presentation. Proceedings from the Credit Scoring and Credit Control XI conference.

Bellotti, T. and Crook, J. 2009. "Macroeconomic conditions in models of Loss Given Default for retail credit." Credit Scoring and Credit Control XI Conference, August.

Benzschawel, T., Haroon, A., and Wu, T. 2011. "A Model for Recovery Value in Default." Journal of Fixed Income, 21(2), 15-29.

Bi, J. and Bennett, K.P. 2003. "Regression error characteristic curves." Proceedings of the Twentieth International Conference on Machine Learning, Washington DC, USA.

Box, G.E.P. and Cox, D.R. 1964. "An analysis of transformations." Journal of the Royal Statistical Society, Series B (Methodological), 26(2), 211-252.

Breiman, L., Friedman, J., Stone, C.J., and Olshen, R.A. 1984. *Classification and Regression Trees.* Chapman & Hall/CRC.

Caselli, S. and Querci, F. 2009. "The sensitivity of the loss given default rate to systematic risk: New empirical evidence on bank loans." Journal of Financial Services Research, 34, 1-34.

Chalupka, R. and Kopecsni, J. 2009. "Modeling Bank Loan LGD of Corporate and SME Segments: A Case Study." Czech Journal of Economics and Finance, 59(4), 360-382

Cohen, J., Cohen, P., West, S. and Aiken, L. 2002. *Applied Multiple Regression/Correlation Analysis for the Behavioral Sciences.* 3rd ed. Lawrence Erlbaum.

Demšar, J. 2006. "Statistical Comparisons of Classifiers over Multiple Data Sets." Journal of Machine Learning Research, 7, 1-30.

Draper, N. and Smith, H. 1998. *Applied Regression Analysis.* 3rd ed. John Wiley.

Fawcett, T. 2006. "An introduction to ROC analysis." Pattern Recognition Letters, 27(8), 861-874.

Freund, R. and Littell, R. 2000. *SAS System for Regression.* 3rd ed. SAS Institute Inc.

Friedman, M. 1940. "A comparison of alternative tests of significance for the problem of m rankings." The Annals of Mathematical Statistics, 11(1), 86-92.

Grunert, J. and Weber, M. 2008. "Recovery rates of commercial lending: Empirical evidence for German companies." Journal of Banking & Finance, 33(3), 505–513.

Gupton, G. and Stein, M. 2002. "LossCalc: Model for predicting loss given default (LGD)." Technical report, Moody's. http://www.defaultrisk.com/_pdf6j4/losscalc_methodology.pdf

Hartmann-Wendels, T. and Honal, M. 2006. "Do economic downturns have an impact on the loss given default of mobile lease contracts? An empirical study for the German leasing market." Working Paper, University of Cologne.

Hlawatsch, S. and Ostrowski, S. 2010. "Simulation and Estimation of Loss Given Default." FEMM Working Papers 100010, Otto-von-Guericke University Magdeburg, Faculty of Economics and Management.

Hlawatsch, S. and Reichling, P. 2010. "A Framework for LGD Validation of Retail Portfolios." Journal of Risk Model Validation, 4(1), 23-48.

Hu, Y.T. and Perraudin, W. (2002). "The dependence of recovery rates and defaults." Mimeo, Birkbeck College.

Jacobs, M. and Karagozoglu, A.K. 2011. "Modeling Ultimate Loss Given Default on Corporate Debt." Journal of Fixed Income, 21(1), 6-20.

Jankowitsch, R., Pillirsch, R., and Veza, T. 2008. "The delivery option in credit default swaps." Journal of Banking and Finance, 32 (7), 1269–1285

Li, H. 2010. "Downturn LGD: A Spot Recovery Approach." MPRA Paper 20010, University Library of Munich, Germany.

Loterman, G., Brown, I., Martens, D., Mues, C., and Baesens, B. 2009. "Benchmarking State-of-the-Art Regression Algorithms for Loss Given Default Modelling." 11th Credit Scoring and Credit Control Conference (CSCC XI). Edinburgh, UK.

Luo, X. and Shevchenko, P.V. 2010. "LGD credit risk model: estimation of capital with parameter uncertainty using MCMC." Quantitative Finance Papers.

Martens, D., Baesens, B., Van Gestel, T., and Vanthienen, J. 2007. "Comprehensible credit scoring models using rule extraction from support vector machines." European Journal of Operational Research, 183(3), 1466-1476.

Martens, D., Baesens, B., and Van Gestel, T. 2009. "Decompositional rule extraction from support vector machines by active learning." IEEE Transactions on Knowledge and Data Engineering, 21(2), 178-191.

Matuszyk, A., Mues, C., and Thomas, L.C. 2010. "Modelling LGD for Unsecured Personal Loans: Decision Tree Approach." Journal of the Operational Research Society, 61(3), 393-398.

Nagelkerke, N.J.D. 1991. "A note on a general definition of the coefficient of determination." Biometrica, 78(3), 691–692.

Qi, M. and Zhao, X. 2011. "Comparison of Modeling Methods for Loss Given Default." Journal of Banking & Finance, 35(11), 2842-2855.

Rosch, D. and Scheule, H. 2008. "Credit losses in economic downtowns – empirical evidence for Hong Kong mortgage loans." HKIMR Working Paper No.15/2008

Shleifer, A. and Vishny, R. 1992. "Liquidation values and debt capacity: A market equilibrium approach." Journal of Finance, 47, 1343-1366.

Sigrist, F. and Stahel, W.A. 2010. "Using The Censored Gamma Distribution for Modeling Fractional Response Variables with an Application to Loss Given Default." Quantitative Finance Papers.

Smithson, M. and Verkuilen, J. 2006. "A better lemon squeezer? Maximum-likelihood regression with beta-distributed dependent variables." Psychological Methods, 11(1), 54-71.

Somers, M. and Whittaker, J. 2007. "Quantile regression for modelling distribution of profit and loss." European Journal of Operational Research, 183(3). 1477-1487,

Van Gestel, T., Baesens, B., Van Dijcke, P., Suykens, J., Garcia, J., and Alderweireld, T. 2005. "Linear and non-linear credit scoring by combining logistic regression and support vector machines." Journal of Credit Risk, 1(4).

Van Gestel, T., Baesens, B., Van Dijcke, P., Garcia, J., Suykens, J. and Vanthienen, J. 2006. "A process model to develop an internal rating system: Sovereign credit ratings." Decision Support Systems, 42(2), 1131-1151.

Van Gestel, T., Martens, D., Baesens, B., Feremans, D., Huysmans, J. and Vanthienen, J. 2007." Forecasting and analyzing insurance companies' ratings." International Journal of Forecasting, 23(3), 513-529.

Van Gestel, T. and Baesens, B. 2009. *Credit Risk Management: Basic Concepts: Financial Risk Components, Rating Analysis, Models, Economic and Regulatory Capital.* Oxford University Press.

Chapter 5 Development of an Exposure at Default (EAD) Model

5.1 Overview of Exposure at Default

Exposure at Default (EAD) can be defined simply as a measure of the monetary exposure should an obligor go into default. Under the Basel II requirements for the advanced internal ratings-based approach (A-IRB), banks must estimate and empirically validate their own models for Exposure at Default (EAD) (Figure 5.1). In practice, however, this is not as simple as it seems, as in order to estimate EAD, for off-balance-sheet (unsecured) items such as example credit cards, one requires the committed but unused loan amount times a credit conversion factor (CCF). Simply setting a CCF value to 1 as a conservative estimate would not suffice, considering that as a borrower's conditions worsen, the borrower typically will borrow more of the available funds.

Note: The term Loan Equivalency Factor (LEQ) can be used interchangeably with the term credit conversion factor (CCF) as CCF is referred to as LEQ in the U.S.

Figure 5.1: IRB and A-IRB Approaches

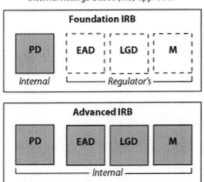

In defining EAD for on-balance sheet items, EAD is typically taken to be the nominal outstanding balance net of any specific provisions (Financial Supervision Authority, UK 2004a, 2004b). For off-balance sheet items (for example, credit cards), EAD is estimated as the current drawn amount, $E(t_r)$, plus the current undrawn amount (credit limit minus drawn amount), $L(t_r) - E(t_r)$, multiplied by a credit conversion factor, CCF or loan equivalency factor (LEQ):

$$\widehat{EAD} = E(t_r) + \widehat{CCF} \times \left(L(t_r) - E(t_r) \right) \quad (5.1)$$

The credit conversion factor can be defined as the percentage rate of undrawn credit lines (UCL) that have yet to be paid out but will be utilized by the borrower by the time the default occurs (Gruber and Parchert, 2006). The calculation of the CCF is required for off-balance sheet items, as the current exposure is generally not a good indication of the final EAD, the reason being that, as an exposure moves towards default, the likelihood is that more will be drawn down on the account. In other words, the source of variability of the exposure is the possibility of additional withdrawals when the limit allows this (Moral, 2006). However, a CCF calculation is not required for secured loans such as mortgages.

In this chapter, a step-by-step process for the estimation of Exposure at Default is given, through the use of SAS Enterprise Miner. At each stage, examples are given using real world financial data. This chapter also develops and computes a series of competing models for predicting Exposure at Default to show the benefits of using the best model. Ordinary least squares (OLS), Binary Logistic and Cumulative Logistic regression models, as well as an OLS with Beta transformation model, are demonstrated to not only show the most appropriate method for estimating the CCF value, but also to show the complexity in implementing each technique. A direct estimation of EAD, using an OLS model, will also be shown, as a comparative measure to first estimating the CCF. This chapter will also show how parameter estimates and comparative statistics can be calculated in Enterprise Miner to determine the best overall model. The first section of this chapter will begin by detailing the potential time horizons you may wish to consider in initially formulating the CCF value. A full description of the data used within this chapter can be found in the appendix section of this book.

5.2 Time Horizons for CCF

In order to initially calculate the CCF value, two time points are required. The actual Exposure at Default (EAD) is measured at the time an account goes into default, but we also require a time point from which the drawn balance and risk drivers can be measured, Δt before default, displayed in Figure 5.2 below:

Figure 5.2: Estimation of Time Horizon

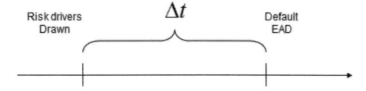

Once we have these two values, the value of CCF can be calculated using the following formulation:

$$CCF_i = \frac{E(t_d) - E(t_r)}{L(t_r) - E(t_r)} \quad (5.2)$$

where $E(t_d)$ is the Exposure at the time of Default, $L(t_r)$ is the advised credit limit at the start of the time period, and $E(t_r)$ is the drawn amount at the start of the cohort. A worked example of the CCF calculation is displayed in Figure 5.3.

Figure 5.3: Example CCF Calculation

Example: $CCF = \dfrac{5500 - 4000}{6000 - 4000} = 0.75$

Drawn amount, at start of cohort = £4000

Limit amount, at start of cohort = £6000

Exposure at default (EAD) = £5500

The problem is how to determine this time period Δt, to select prior to the time of default. To achieve this, three types of approach that can be used in the selection of the time period Δt for calculating the credit conversion factor have been proposed:

1. The Cohort Approach (Figure 5.4) – This approach groups defaulted accounts into discrete calendar periods according to the date of default. A common length of time for these calendar periods is 12 months; however, shorter time periods maybe more appropriate if a more conservative approach is required. The information for the risk drivers and drawn/undrawn amounts are then collected at the start of the calendar period along with the drawn amount at the actual time of default (EAD). With the separation of data into discrete cohorts, the data can then be pooled for estimation. An example of the cohort approach can be seen in the following diagram. For example, calendar period is defined as 1st November 2002 to 30th October 2003. The information about the risk drivers, drawn/undrawn amounts on the 1st November 2002 are then collected as well as the drawn amounts at the time of any accounts going into default during that period.

Figure 5.4: Cohort Approach

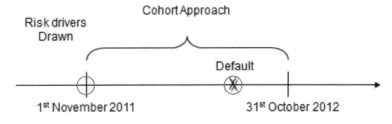

2. The Fixed-Horizon Approach (Figure 5.5) – For this approach, information regarding risk drivers and drawn/undrawn amounts is collected at a fixed time period prior to the defaulting date of a facility as well as the drawn amount on the date of default. In practice, this period is usually set to 12 months unless other time periods are more appropriate or conservative. For example, if a default were to occur on the 15th May 2012, then the information about the risk drivers and drawn/undrawn amount of the defaulted facility would be collected from 15th May 2011.

Figure 5.5: Fixed-Horizon Approach

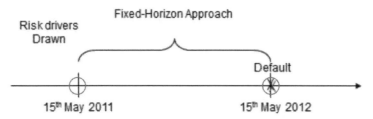

3. The Variable Time Horizon Approach (Figure 5.6) – This approach is a variation of the fixed-horizon approach by first fixing a range of horizon values (12 months) in which the CCF will be computed. Second, the CCF values are computed for each defaulted facility associated with a set of reference dates (1 month, 2 months, …, 12 months before default). Through this process, a broader set of potential default dates are taken into consideration when estimating a suitable value for the CCF. An example of this is shown in Figure 5.6.

Figure 5.6: Variable Time Horizon Approach

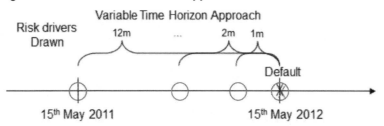

As to which is the most appropriate time horizon to use in the initial calculation of the CCF value, this is very much a matter of business knowledge of the portfolio you are working with and, to an extent, personal preference. With regards to commonality, the Cohort Approach is widely used in the formulation process of the CCF value; hence, in the model build process of this chapter this approach will be used in calculating the CCF.

5.3 Data Preparation

The example data set used to demonstrate the development of an Exposure at Default Model contains 38 potential input variables, an ID variable, and the target variable. As with any model development, one must begin with the data, implementing both business knowledge as well as analytical techniques to determine the robustness of the data available.

Here, a default is defined to have occurred on a credit card when a charge off has been made on that account (a charge off in this case is defined as the declaration by the creditor that an amount of debt is unlikely to be collected, declared at the point of 180 days or 6 months without payment). In order to calculate the CCF value, the original data set has been split into two 12-month cohorts, with the first cohort running from November 2002 to October 2003 and the second cohort from November 2003 to October 2004 (Figure 5.7). As explained in the previous section, the cohort approach groups defaulted facilities into discrete calendar periods, in this case 12-month periods, according to the date of default. Information is then collected regarding risk factors and drawn/undrawn amounts at the beginning of the calendar period and drawn amount at the date of default. The cohorts have been chosen to begin in November and end in October in order to reduce the effects of any seasonality on the calculation of the CCF.

Figure 5.7: Enterprise Miner Data Extract

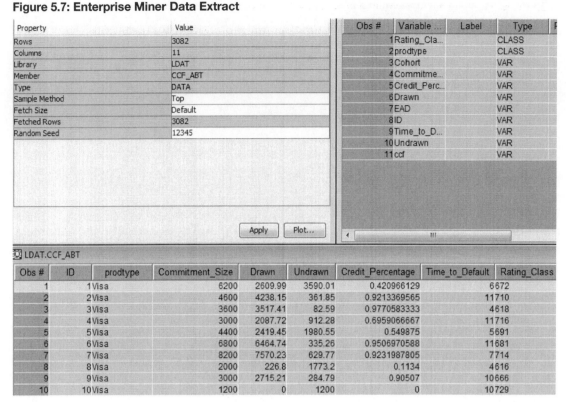

The characteristics of the cohorts used in evaluating the performance of the regression mode are given below in Table 5.1:

Table 5.1: Characteristics of Cohorts for EAD Data Set

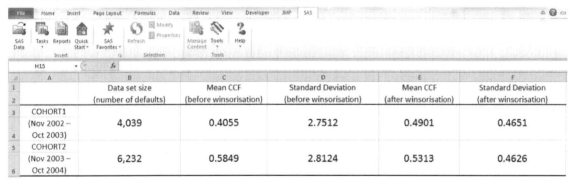

	Data set size (number of defaults)	Mean CCF (before winsorisation)	Standard Deviation (before winsorisation)	Mean CCF (after winsorisation)	Standard Deviation (after winsorisation)
COHORT1 (Nov 2002 – Oct 2003)	4,039	0.4055	2.7512	0.4901	0.4651
COHORT2 (Nov 2003 – Oct 2004)	6,232	0.5849	2.8124	0.5313	0.4626

In our Enterprise Miner examples, COHORT1 will be used to train the regression models, while COHORT2 will be used to test the performance of the model (out-of-time testing) (following Figure 5.8).

Figure 5.8: Enterprise Miner Data Nodes

Both data sets contain variables detailing the type of defaulted credit card product and the following monthly variables: advised credit limit, current balance, the number of days delinquent, and the behavioral score.

The following variables are also required in the computation of a CCF value based on the monthly data found in each of the cohorts, where t_d is the default date and t_r is the reference date (the start of the cohort):

- Committed amount, $L(t_r)$: the advised credit limit at the start of the cohort;

- Drawn amount, $E(t_r)$: the exposure at the start of the cohort;

- Undrawn amount, $L(t_r) - E(t_r)$: the advised limit minus the exposure at the start of cohort;

- Credit percentage usage $\dfrac{E(t_r)}{L(t_r)}$: the exposure at the start of the cohort divided by the advised credit limit at the start of the cohort;

- Time to default, $(t_d - t_r)$: the default date minus the reference date (in months);

- Rating class, $R(t_r)$: the behavioral score at the start of the cohort, binned into four discrete categories 1: AAA-A; 2: BBB-B; 3: C; 4: UR (unrated).

The CCF variable itself can then be computed as follows:

- Credit conversion factor, CCF_i: calculated as the actual EAD minus the drawn amount at the start of the cohort divided by the advised credit limit at the start of the cohort minus the drawn amount at the start of the cohort:

$$CCF_i = \frac{E(t_d) - E(t_r)}{L(t_r) - E(t_r)} \quad (5.3)$$

In addition to the aforementioned variables, a set of additional variables that could potentially increase the predictive power of the regression models implemented can be used. These additional variables of use are:

- Average number of days delinquent in the previous 3 months, 6 months, 9 months, and 12 months. It is expected that the higher the number of days delinquent closer to default date, the higher the CCF value will be;

- Increase in committed amount: binary variable indicating whether there has been an increase in the committed amount since 12 months prior to the start of the cohort. It is expected that an increase in the committed amount increases the value of the CCF;

- Undrawn percentage, $\dfrac{L(t_r) - E(t_r)}{L(t_r)}$: the undrawn amount at the start of the cohort divided by the advised credit limit at the start of the cohort. It is expected that higher ratios result in a decrease in the value of the CCF;

- Absolute change in drawn, undrawn, and committed amount: variable amount at t_r minus the variable amount 3 months, 6 months, or 12 months prior to t_r;

- Relative change in drawn, undrawn, and committed amount: variable amount at t_r minus the variable amount 3 months, 6 months, or 12 months prior to t_r, divided by the variable amount 3 months, 6 months, or 12 months prior to t_r, respectively.

The potential predictiveness of all the variables proposed in this chapter will be evaluated by calculating the information value (IV) based on their ability to separate the CCF value into either of two classes, $0: CCF < \overline{CCF}$ (non-event), and $1: CCF \geq \overline{CCF}$ (event).

After binning input variables using an entropy-based procedure, implemented in the **Interactive Grouping node** (Credit Scoring tab) in SAS Enterprise Miner, the information value of a variable with k bins is given by:

$$IV = \sum_{i=1}^{k} \left[\left(\frac{n_1(i)}{N_1} - \frac{n_0(i)}{N_0} \right) \ln \left(\frac{n_1(i)/N_1}{n_0(i)/N_0} \right) \right] \quad (5.4)$$

where $n_0(i), n_1(i)$ denotes the number of non-events (non-default) and events (default) in bin i, and N_0, N_1 are the total number of non-events and events in the data set, respectively.

This measure allows us to do a preliminary screening of the relative potential contribution of each variable in the prediction of the CCF.

The distribution of the raw CCF for the first Cohort (COHORT1) is shown here:

Figure 5.9: CCF Distribution (Scale -10 to +10 with Point Distribution Around 0 and 1)

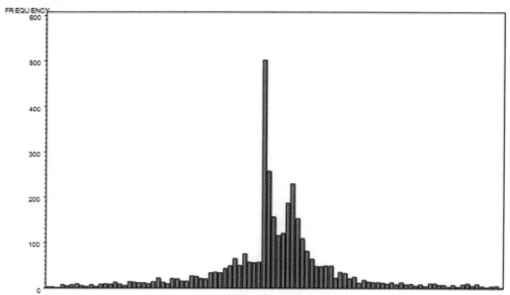

The raw CCF displays a substantial peak around 0 and a slight peak at 1 with substantial tails either side of these points. Figure 5.9 displays a snapshot of CCF values in the period -10 to 10. This snapshot boundary has been selected to allow for the visualization of the CCF distribution. Values of $CCF > 1$ can occur when the actual EAD is greater than the advised credit limit, whereas values of $CCF < 0$ can occur when both the drawn amount and the EAD exceed the advised credit limit or where the EAD is smaller than the drawn amount. In practice, this occurs as the advised credit limit and drawn amount are measured at a time period, t_r, prior to default, and therefore at t_d the advised credit limit maybe higher or lower than at l_r. Extremely large positive and negative values of CCF can also occur if the drawn amount is slightly above or below the advised credit limit:

$$CCF_i = \frac{E(t_d) - E(t_r)}{L(t_r) - E(t_r)} = \frac{3300 - 3099.9}{3100 - 3099.9} = 2001 \quad (5.5)$$

$$CCF_i = \frac{E(t_d) - E(t_r)}{L(t_r) - E(t_r)} = \frac{3000 - 3500}{4000 - 3500} = -1 \quad (5.6)$$

It therefore seems reasonable to winsorise the data so that the CCF can only fall between values of 0 and 1. The following Figure 5.10 displays the same CCF value winsorised at 0 and 1:

Figure 5.10: CCF Winsorised Distribution

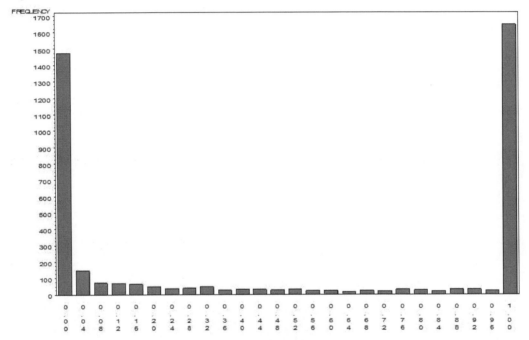

The winsorised CCF yields a bimodal distribution with peaks at 0 and 1, and a relatively flat distribution between the two peaks. This bears a strong resemblance to the distributions identified in loss given default modeling (LGD) (Matuszyk et al., 2010). In the estimation of the CCF it is common industry practice to use this limited CCF between 0 and 1, as it retains the realistic bounds of the CCF value. In order to create this winsorisation a **SAS Code node** (Truncation) is used (Figure 5.11). Figure 5.12 details the code utilized to achieve this, where tgt_var is equal to the CCF variable.

Figure 5.11: EM Transformation Process Flow

Figure 5.12: Data step code to winsorise the CCF

Training Code

```
DATA &EM_EXPORT_TRAIN;
    SET &EM_IMPORT_DATA;
        IF TGT_VAR LE 0 THEN TGT_VAR=0;
        IF TGT_VAR GE 1 THEN TGT_VAR=1;
RUN;
```

5.4 CCF Distribution – Transformations

A common feature of the CCF value displayed in the majority of credit portfolios is that the CCF value exhibits a bi-modal distribution with two peaks around 0 and 1, and a relatively flat distribution between those peaks. This non-normal distribution is therefore less suitable for modeling with traditional ordinary least squares (OLS) regression. The motivation for using an OLS with Beta transformation model is that it accounts for a range of distributions including a U-shaped distribution. A direct OLS estimation of the EAD value is another potential methodology for estimating EAD; however, regulators require an estimation of CCF to be made where credit lines are revolving. This model build process will also be shown later in this chapter.

Outlier detection is applied using the **Filter node** utilizing the methodologies detailed in Chapter 2 to remove extreme percentile values for the independent variable. Missing value imputation is also applied using a tree-based approach for categorical and interval variables to best approximate those input variables where values do not exist. In order to achieve this beta transformation, we must first introduce a **SAS Code node** to our process flow (Figure 5.13):

Figure 5.13: Enterprise Miner Process Flow Including Truncation, Outlier Detection, Imputation and Beta-Normal Transformation

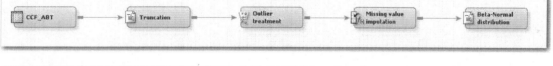

Training Code

```
DATA TEMP(KEEP=tgt_var);
  SET &EM_IMPORT_DATA(KEEP=tgt_var);
RUN;
PROC UNIVARIATE DATA=TEMP;
  PROBPLOT tgt_var/ BETA(ALPHA=EST BETA=EST theta = -0.001);
  OUTPUT OUT=&em_lib..ESTIMATE MEAN=MU STD=SD;
RUN;

DATA &em_lib..PARMS;
  SET &em_lib..ESTIMATE;
  ALPHA = (MU**2*(1-MU))/(SD**2) - MU;
  BETA = ALPHA * ((1/MU) -1);
RUN;

DATA &EM_EXPORT_TRAIN;
    IF _N_=1 THEN SET &em_lib..PARMS;
    SET &EM_IMPORT_DATA;
        CCF = PROBIT(CDF('BETA',tgt_var, ALPHA, BETA));
RUN;
```

Within this **SAS Code node,** we can apply a transformation to the tgt_var using the cumulative distribution function (CDF).

Whereas OLS regression tests generally assume normality of the dependent variable y, the empirical distribution of CCF can often be approximated more accurately by a beta distribution. Assuming that y is constrained to the open interval $(0,1)$, the cumulative distribution function (CDF) of a beta distribution is given by:

$$\beta(y;a,b) = \frac{\Gamma(a+b)}{\Gamma(a)\Gamma(b)} \int_0^y v^{a-1}(1-v)^{b-1}\, dv \quad (5.7)$$

where $\Gamma()$ denotes the well-known gamma function, and a and b are two shape parameters, which can be estimated from the sample mean μ and variance σ^2 using the method of the moments:

$$a = \frac{\mu^2(1-\mu)}{\sigma^2} - \mu\,; \qquad b = a\left(\frac{1}{\mu} - 1\right) \quad (5.8)$$

A potential solution to improve model fit therefore is to estimate an OLS model for a transformed dependent variable $y_i^* = N^{-1}\big(\beta(y_i;a,b)\big)\ (i=1,\ldots,l)$, in which $N^{-1}()$ denotes the inverse of the standard normal CDF.

The Beta-Normal distribution **SAS Code node** is connected to a **Data Partition node** to create 70% Train and 30% Validation samples. The **Metadata node** is then used to change the role of target to the beta-transformed CCF value prior to an OLS regression model being fitted. The predictions by the OLS model are then transformed back through the standard normal CDF and the inverse of the fitted Beta CDF to get the actual CCF estimates. To achieve this in SAS Enterprise Miner the following code can be used in a **SAS Code node** to transform the CCF values back to the original tgt_var variable displayed in Figure 5.14 below.

Figure 5.14: Enterprise Miner Process Flow Inversion of Beta-Normal Transformation

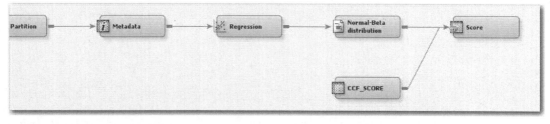

```
Training Code

proc sql noprint;
    select ALPHA into :A from &em_lib..ESTIMATE1;
    select BETA into :B from &em_lib..ESTIMATE1;
quit;
DATA &EM_EXPORT_TRAIN;
SET &EM_IMPORT_DATA;
tgt_var = QUANTILE ('beta', Probnorm(CCF),&A,&B);
RUN;
```

5.5 Model Development

5.5.1 Input Selection

In this section, an analysis of the input variables and their relationship to the dichotomized CCF value ($0 : CCF < \overline{CCF}$; $1 : CCF \geq \overline{CCF}$) will be looked at. The following table displays the resulting information value for the top 10 variables, ranked from most to least predictive:

Table 5.2: Information Values of Constructed Variables

Variable	Information Value
Credit percentage usage	1.825
Undrawn percentage	1.825
Undrawn	1.581
Relative change in undrawn amount (12 months)	0.696
Relative change in undrawn amount (6 months)	0.425
Relative change in undrawn amount (3 months)	0.343
Rating Class	0.233
Time-to-Default	0.226
Drawn	0.181
Absolute change in drawn amount (3 months)	0.114

Typically, input variables which display an information value greater than 0.1 are deemed to have a significant contribution in the prediction of the target variable. From this analysis, it is shown that the majority of the relative and absolute changes in drawn, undrawn, and committed amounts do not possess the same ability to discriminate between low and high CCFs as the original variable measures at only reference time. It is also discernable from the results that the undrawn amount could be an important variable in the discrimination of the CCF value. It must be taken into consideration, however, that although the variables may display a good ability to discriminate between the low and high CCFs, the variables themselves are highly correlated with each other. This is an important consideration in the creation of CCF models, as although we want to consider as many variables as possible, variable interactions can skew the true relationship with the target.

5.5.2 Model Methodology

OVERVIEW OF TECHNIQUES

In this section, we look at the types of techniques and methodologies that can be applied in the estimation of the CCF value. It is important to understand the definition of each technique and how it can be utilized to solve the particular problems discussed in this chapter. As such, we begin by defining the formulation of the technique and go on to show how this can be implemented within SAS/STAT.

ORDINARY LEAST SQUARES

Ordinary least squares regression (see Draper and Smith, 1998) is probably the most common technique to find the optimal parameters $\mathbf{b}^T = [b_0, b_1, b_2, ..., b_n]$ to fit the following linear model to a data set:

$$y = \mathbf{b}^T \mathbf{x}^T \quad (5.9)$$

where $\mathbf{x}^T = [1, x_1, x_2, ..., x_n]$. OLS solves this problem by minimizing the sum of squared residuals which leads to:

$$\sum_{i=1}^{l} (e_i)^2 = \sum_{i=1}^{l} (y_i - \mathbf{b}^T \mathbf{x}_i)^2 \quad (5.10)$$

with $\mathbf{x}^T = [x_1, x_2, \ldots, x_l]$ and $\mathbf{y} = [y_1, y_2, \ldots, y_l]^T$.

The SAS code used to calculate the OLS regression model is displayed in the following Figure 5.15:

Figure 5.15: SAS/STAT code for Regression Model Development

```
PROC REG DATA = Cohort1 OUTEST = out RSQUARE;
    MODEL ccf = {Rating_Grade1 Rating_Grade2 Rating_Grade3 Rating_Grade4}
    &inputs /SELECTION = stepwise SLENTRY =0.01 SLSTAY = 0.01 GROUPNAMES = 'Dummy for Rating Grade';
    OUTPUT OUT = t STUDENT = res COOKD = cookd PREDICTED = parms;
RUN;
QUIT;
```

The &inputs statement refers to a %let macro containing a list of all the input variables calculated in the estimation of the CCF. These variables are detailed in the following EMPIRICAL SET-UP AND DATA section.

BINARY AND CUMULATIVE LOGIT MODELS

The CCF distribution is often characterized by a peak around CCF = 0 and a further peak around CCF = 1 (Figure 5.9 and Figure 5.10). This non-normal distribution can lead to inaccurate linear regression models. Therefore, it is proposed that binary and cumulative logit models can be used in an attempt to resolve this issue by grouping the observations for the CCF into two categories for the binary logit model and three categories for the cumulative logit model. For the binary response variable, two different splits will be tried: the first is made according to the mean of the CCF distribution (Class $0 : CCF < \overline{CCF}$; Class $1: CCF \geq \overline{CCF}$) and the second is made based on whether the CCF is less than 1 (Class $0 : CCF < 1$, Class $1: CCF \geq 1$). For the cumulative logit model, the CCF is split into three levels: Class $0 : CCF = 0$, Class $1: 0 < CCF < 1$ and Class $2 : CCF = 1$.

For the binary logit model (see Hosmer and Stanley, 2000), a sigmoid relationship between $P(\text{class} = 1)$ and $\mathbf{b}^T\mathbf{x}$ is assumed such that $P(\text{class} = 1)$ cannot fall below 0 or above 1:

$$P(\text{class} = 1) = \frac{1}{1+e^{-(\mathbf{b}^T\mathbf{x})}} \quad (5.11)$$

The SAS code used to calculate the binary logit model is displayed in Figure 5.16:

Figure 5.16: SAS/STAT code for Logistic Regression Model Development

```
PROC LOGISTIC DATA=Cohort1 des outmodel=param out=out;
  CLASS rating_grade;
  MODEL ccf_bin = &inputs Rating_Grade /RSQUARE SELECTION=stepwise SLENTRY=0.01
    SLSTAY=0.01 STB;
  OUTPUT PRED=lpredy;
  score DATA=Cohort2 out=scored outroc=roc;
RUN;
```

The cumulative logit model is simply an extension of the binary two-class logit model which allows for an ordered discrete outcome with more than 2 levels $(k > 2)$:

$$P(\text{class} \leq j) = \frac{1}{1+e^{-(d_j + b_1 x_1 + b_2 x_2 + \ldots + b_n x_n)}} \quad (5.12)$$

$$j = 1, 2, \ldots, k-1$$

The cumulative probability, denoted by $P(\text{class} \leq j)$, refers to the sum of the probabilities for the occurrence of response levels up to and including the jth level of y. The SAS code for the cumulative logit model is a variant on the binary logit coding with the use of a link=CLOGIT in the **proc logistic** model statement.

5.5.3 Performance Metrics

When developing any model, it is important to determine how well the model built performs on an out-of-sample or hold-out-sample data set. Performance metrics that are commonly used to evaluate credit risk models are the r-square value, the root mean squared error value, and correlation coefficients (Pearson's r and Spearman's p). These performance metrics can be computed using a **SAS Code node** after the **Score node**.

A description of each of the performance metrics is detailed as follows:

R-Square

The R-square formulation is discussed in Chapter 4. In order to calculate the performance metrics on the categorical predictions made by the LOGIT and CLOGIT models, first a continuous prediction value must be obtained. This is achieved by multiplying the probability of being in each of the bins by the average CCF value for each of those respective bins and summing the result, thus obtaining an expected value of CCF. After this value has been computed, the resulting value is then used in the calculation of the performance metrics.

Pearson's Correlation Coefficient

Please refer to Chapter 4.

Spearman's Correlation Coefficient

Please refer to Chapter 4.

Root Mean Squared Error

Please refer to Chapter 4.

Subsequently, the performance of the models themselves in the prediction of the CCF is examined. Table 5.3 reports the parameter estimates and p-values for the variables used by each of the regression techniques implemented. The parameter signs found in Jacobs are also shown for comparative purposes (2008). The five regression models detailed are: an OLS model implementing only the suggested predictive variables in Moral, (2006); an OLS model incorporating the additional variables after stepwise selection; an OLS with Beta transformation model; a binary logit model; and a cumulative logit model. For the binary logit model, the best class split found was to select $0:CCF<1$ and $1:CCF\geq1$. It is, however, important to note that little difference was found between the choices of class split for the binary model.

From Table 5.3, it can be seen that the best performing regression algorithm for all three performance measures is the binary logit model with an R^2 value of 0.1028. Although this R^2 value is low, it is comparable to the range of performance results previously reported in other work on LGD modeling (see Chapter 5). This result also re-affirms the proposed usefulness of a logit model for estimating CCFs in Valvonis (2008). It can also be seen that all five models are quite similar in terms of variable significance levels and positive/negative signs. There does, however, seem to be some discrepancy for the Rating class variable, where the medium-range behavioral score band appears to be associated with the highest CCFs.

Table 5.3: Parameter Estimates and P-Values for CCF Estimation on the COHORT2 Data Set

Variables	Coefficient sign reported in Jacobs, (2008)	OLS model (using only suggested variables in Moral, (2006))		OLS model (OLS) (additional variables)	
		Parameter Estimate	P-value	Parameter Estimate	P-value
Intercept 1		0.1830	<.0001	0.1365	<.001
Intercept 2					
Credit percentage usage	−	-0.1220	<.001	-0.1260	<.001
Committed amount	+	1.73E-05	<.0001	1.76E-05	<.0001
Undrawn	+	-8.68E-05	<.0001	-8.88E-05	<.0001
Time-to-Default	+	0.0334	<.0001	0.0326	<.0001
Rating class	−				
Rating 1 (AAA-A) vs. 4 (UR)		0.1735	<.0001	0.2304	<.0001
Rating 2 (BBB-B) vs. 4 (UR)		0.2483	<.0001	0.2977	<.0001
Rating 3 (C) vs. 4 (UR)		0.0944	<.0001	0.1201	<.0001
Average number of days delinquent in the last 6 months	N/A			0.0048	<.0001
Undrawn percentage	N/A				

Table 5.3: Parameter Estimates and P-Values for CCF Estimation on the COHORT2 Data Set

Variables	OLS with Beta transformation (B-OLS)		Binary logit model (LOGIT)		Cumulative logit model (CLOGIT)	
	Parameter Estimate	P-value	Parameter Estimate	P-value	Parameter Estimate	P-value
Intercept 1	-0.5573	<.0001	-1.5701	<.0001	0.6493	<.0001
Intercept 2					-0.5491	<.001
Credit percentage usage			-0.5737	<.001	-1.3220	<.0001
Committed amount	2.2E-05	<.0001	9.0E-05	<.0001	8.8E-05	<.0001
Undrawn	-1.1E-04	<.0001	-4.7E-04	<.0001	-3.6E-04	<.0001
Time-to-Default	0.0358	<.0001	0.1538	<.0001	0.1009	<.0001
Rating class						
Rating 1 (AAA-A) vs. 4 (UR)	0.2223	<.0001	0.4000	0.0069	-0.0772	0.5472
Rating 2 (BBB-B) vs. 4 (UR)	0.3894	<.0001	0.5885	<.0001	0.6922	<.0001
Rating 3 (C) vs. 4 (UR)	0.1664	<.0001	-0.2121	0.0043	-0.0157	0.8098
Average number of days delinquent in the last 6 months	0.0062	<.0001	0.0216	<.0001	0.0218	<.0001
Undrawn percentage	0.2784	<.0001				

Table 5.4: Performance Metrics for CCF Estimation on the COHORT2 Data Set

	OLS model (using only suggested variables in Moral, (2006))	OLS model (OLS) (additional variables)	OLS with Beta transformation (B-OLS)
R-square (R^2)	0.0982	0.0960	0.0830
Pearson's Correlation (r)	0.3170	0.3144	0.3125
Spearman's Correlation (ρ)	0.2932	0.2943	0.3000
Root Mean Squared Error (RMSE)	0.4393	0.4398	0.4433

Table 5.4: Performance Metrics for CCF Estimation on the COHORT2 Data Set

	Binary logit model (LOGIT)	Cumulative logit model (CLOGIT)
R-square (R^2)	0.1028	0.0822
Pearson's Correlation (r)	0.3244	0.2897
Spearman's Correlation (ρ)	0.3283	0.2943
Root Mean Squared Error (RMSE)	0.4704	0.4432

Of the additional variables tested (absolute or relative change in the drawn amount, credit limit, and undrawn amount), only 'Average number of days delinquent in the last 6 months' and 'Undrawn percentage' were retained by the stepwise selection procedures. This is most likely due to the fact that their relation to the CCF is already largely accounted for by the base model variables. It is also of interest to note that although one additional variable is selected in the stepwise procedure for the second OLS model, there is no increase in predictive power over the original OLS model.

A direct estimation of the un-winsorised CCF with the use of an OLS model was also undertaken. The results from this experimentation indicate that it is even harder to predict the un-winsorised CCF than the CCF winsorised between 0 and 1 with a predictive performance far weaker than the winsorised model. When these results are applied to the estimation of the actual EAD an inferior result is also achieved.

With the predicted values for the CCF obtained from the five models, it is then possible to estimate the actual EAD value for each observation i in the COHORT2 data set, as follows:

$$\widehat{EAD} = E(t_r) + \widehat{CCF}.(L(t_r) - E(t_r)) \quad (5.13)$$

This gives us an estimated "monetary EAD" value which can be compared to the actual EAD value found in the data set. For comparison purposes, a conservative estimate for the EAD (assuming CCF = 1) is also calculated, as well as an estimate for EAD where the mean of the CCF in the first cohort is used (Table 5.5). The following table (Table 5.6) displays the predictive performance of this estimated EAD amount against the actual EAD amount:

Table 5.5: EAD Estimates Based on Conservative and Mean Estimate for CCF

	Conservative estimate of EAD (CCF=1)	Estimate of EAD where CCF equals the mean CCF in first cohort
R-square (R2)	0.5178	0.6486
Pearson's Correlation (r)	0.7588	0.8062
Spearman's Correlation (ρ)	0.6867	0.7354

Table 5.6: EAD Estimates Based on CCF Predictions Against Actual EAD Amounts

	OLS model (using only previously suggested variables)	OLS model (including average number of days delinquent in the last 6 months)	OLS with Beta transformation (B-OLS)	Binary logit model (LOGIT)	Cumulative logit model (CLOGIT)
R-square	0.6450	0.6431	0.8365	0.6344	0.6498
Pearson's Correlation	0.8049	0.8038	0.8000	0.8016	0.8068
Spearman's Correlation	0.7421	0.7405	0.7270	0.7387	0.7381

It can be determined from these results that although the predicted CCF value gave a relatively weak performance, when this value is applied to the calculation of the estimated EAD formulation, a significant improvement over the conservative model can be made. It can also be noted that the application of the OLS with Beta transformation model gives a significantly higher value for the R-square (0.8365), although the correlation values are comparative to the other models. A possible reason for this is that even though the CCF has been winsorised prior to estimation, the B-OLS model's predictions are much closer to the real CCF values before winsorisation. Thus the B-OLS model produces a better actual estimate of the EAD. However, by simply applying the mean of the CCF, a similar result to the other predicted models can be achieved.

The direct estimation of the EAD, through the use of an OLS model, has also been taken into consideration, without the first estimation and application of a CCF. The results from this direct estimation of EAD are shown in Table 5.7, with the distribution for the direct estimation of EAD given in Figure 5.17.

Table 5.7: Direct Estimation of EAD

Variables	OLS model (direct estimation of EAD)
R-square (R^2)	0.6608
Pearson's Correlation (r)	0.8130
Spearman's Correlation (ρ)	0.7493

It is self-evident from the performance metrics and the produced distribution that a direct estimation of EAD without firstly estimating and applying a CCF can indeed produce reasonable estimations for the actual EAD. This goes someway in ratifying the findings show by Taplin et al. (2007).

Figure 5.17: Distribution of Direct Estimation of EAD

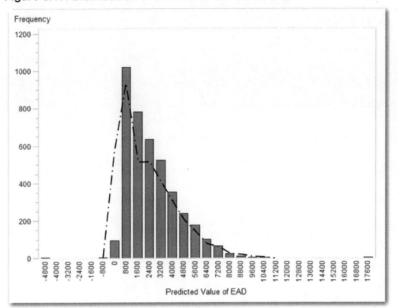

The legend for Figure 5.17 details the frequency of values along the y-axis and the estimated EAD value along the x-axis. The actual EAD amount present is indicated by the overlaid black line.

5.6 Model Validation and Reporting

5.6.1 Model Validation

After the final CCF model has been selected, the modeling results need to be validated. As highlighted in the data preparation section, in order to validate the created model, a hold-out sample or validation sample must be used. This is data that has not been used in the model-building process and thus allows us to validate the model we have built on an unseen data set. For the demonstration in this book, we have split the initial data sample into a 70% modeling data set (training set) and a 30% hold out sample (validation set). In practice, however, it may make more sense to choose different time periods for your training and validation sets so that your model takes other time-related factors into consideration.

The purpose of validation is to check the strength of the development sample and to also ensure that the model has not been over-fitted, which would give you a false indication to its performance power. The way to achieve this in SAS EM is to compare the distributions and performance metrics on the validation data set to that of the training data set. The model built is then validated if the results from each data set are not significantly different. The results from the **Model Comparison node** automatically display graphs (such as an ROC plot, Figure 5.18) and statistics for analysts to determine how well their models are performing.

Figure 5.18: ROC Validation

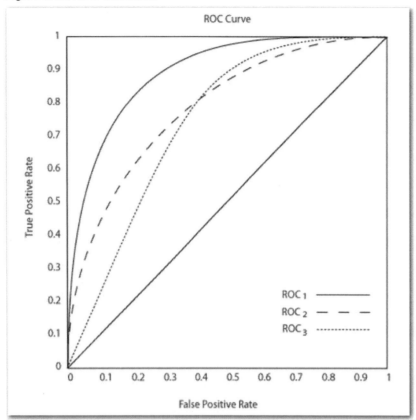

A second way to validate is to compare development statistics for the training and validation samples. The model is validated when there is no significant difference between the statistics for the two samples.

5.6.2 Reports

After the model is validated and the final model is selected, a complete set of reports needs to be created. Reports include information such as model performance measures, development scores, and model characteristics distributions, expected bad/approval rate charts, and the effects of the model on key subpopulations. Reports help make operational decisions such as deciding the model cutoff, designing account acquisition and management strategies, and monitoring models.

Some of the information in the reports is generated using SAS Enterprise Miner and is described below.

Table 5.8: Variable Worth Statistics

Variable	Information Value
Credit percentage usage	1.825
Undrawn percentage	1.825
Undrawn	1.581
Relative change in undrawn amount (12 months)	0.696

For more information about the Gini Statistic and Information Value, see Chapter 3.

Strength Statistics

Another way to evaluate the model are by using the strength statistics: Kolmogorov-Smirnov, Area Under the ROC, and Gini Coefficient.

Table 5.9: Model Strength Statistics

Statistic	Value
KS Statistic	0.481
AUC	0.850
Gini	0.700

Kolmogorov-Smirnov measures the maximum vertical separation between the cumulative distributions of good applicants and bad applicants. The Kolmogorov-Smirnov statistic provides the vertical separation that is measured at one point, not on the entire score range. A model that provides better prediction has a smaller value for the Kolmogorov-Smirnov statistic.

Area Under ROC measures the predictive power across the entire score range by calculating the area under the Sensitivity versus (1-Specificity) curve for the entire score range. The Area Under ROC statistic usually provides a better measure of the overall model strength. A model that is better than random has a value of greater than 0.5 for the Area Under ROC.

Gini Coefficient measures the predictive power of the model. A model that provides better prediction has a smaller value for the Gini Coefficient.

Model Performance Measures

Lift

Lift charts (Figure 5.19) help measure model performance. Lift charts have a lift curve and a baseline. The baseline reflects the effectiveness when no model is used, and the lift curve reflects the effectiveness when the predictive model is used. A greater area between the lift curve and the baseline indicates a better model.

Figure 5.19: Model Lift Chart

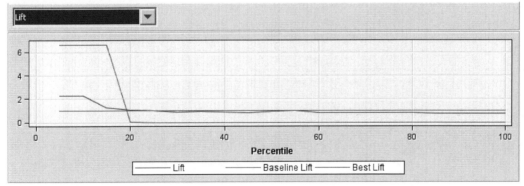

For more information about model performance measures, see the online Help for SAS Enterprise Miner.

Other Measures

Other Basel II model measures that may wish to be calculated from the model to satisfy the Basel II back testing criteria are:

- Pietra Index
- Bayesian Error Rate

- Distance Statistics
- Miscellaneous Statistics
- Brier Score
- Kendall's Tau-b
- Somers' D
- Hosmer-Lemeshow Test
- Observed versus Estimated Index
- Graph of Fraction of Events Versus Fraction of Non-Events
- Binomial Test
- Traffic Lights Test
- Confusion Matrix and Related Statistics
- System Stability Index
- Confidence Interval Tests

Tuning the Model

There are two general approaches to the model-tuning process.

- Refresh the model periodically (for example, each month).
- Refresh the model when needed (based on specific events).

For example, monitoring the selected accuracy measure (Captured Event Rate, Lift, and so on) can signal the need to refresh the analytical model. Monitoring includes calculating the measure and verifying how the value of the measure changes over time. If the value of the measure drops below a selected threshold value, the analytical model must be refreshed.

These monitoring approaches can be achieved through SAS Enterprise Guide where users will define their own KPI's and monitoring metrics or through SAS Model Manager which provides users with industry standard monitoring reports.

5.7 Chapter Summary

In this chapter, the processes and best-practices for the development of an Exposure at Default model using SAS Enterprise Miner and SAS/STAT have been given.

A full development of comprehensible and robust regression models for the estimation of Exposure at Default (EAD) for consumer credit through the prediction of the credit conversion factor (CCF) has been detailed. An in-depth analysis of the predictive variables used in the modeling of the CCF has also been given, showing that previously acknowledged variables are significant and identifying a series of additional variables.

As the model build shows, a marginal improvement in the R-square can be achieved with the use of a binary logit model over a traditional OLS model. Interestingly, the use of a cumulative logit model performs worse than both the binary logit and OLS models. The probable cause of this are the size of the peaks around 0 and 1, compared to the number of observations found in the interval between the two peaks. This, therefore, allows for more error in the prediction of the CCF via a cumulative three-class model.

Another interesting point to note is that although the predictive power of the CCF is weak, when this predicted value is applied to the EAD formulation to predict the actual EAD value, the predictive power is fairly strong. In particular when the predictive values obtained through the application of the OLS with Beta transformation model were applied to the EAD formulation, an improvement in the R-square was seen. Nonetheless, similar performance, in terms of correlations, could be achieved by a simple model that takes the average CCF of the previous cohort, showing that much of the explanatory power of EAD modeling derives from the current exposure.

With regards to the additional variables analyzed in the prediction of the CCF only one (the average number of days delinquent in the last 6 months) gave an adequate p-value, whilst undrawn percentage, potentially an alternative to credit percentage, was significant for the OLS with Beta transformation model. Even though the relative changes in the undrawn amount give reasonable information value scores, these variables do not prove to be significant in the regression models, probably due to their high correlation with the undrawn variable. This shows that the actual values at the start of the cohort already give a significant representation of previous activity in order to predict the CCF.

5.8 References and Further Reading

Draper, N., and Smith, H. 1998. *Applied Regression Analysis.* 3rd ed. John Wiley.

Financial Services Authority, UK. 2004a. "Issues arising from policy visits on exposure at default in large corporate and mid market portfolios." Working Paper, September. http://www.fsa.gov.uk/pubs/international/crsg_visits_portfolios.pdf

Financial Supervision Authority, UK. 2007. "Own estimates of exposure at default." Working Paper, November.

Gruber, W. and Parchert, R. 2006. "Overview of EAD estimation concepts," in: Engelmann B, Rauhmeier R (Eds), The Basel II Risk Parameters: Estimation, Validation and Stress Testing, Springer, Berlin, 177-196.

Hosmer, D.W. and Lemeshow, L. 2000. *Applied Logistic Regression.* 2nd ed. New York; Chichester, John Wiley.

Jacobs, M. 2008. "An Empirical Study of Exposure at Default." OCC Working Paper. Washington, DC: Office of the Comptroller of the Currency.

Matuszyk, A., Mues, C., and Thomas, L.C. 2010. "Modelling LGD for Unsecured Personal Loans: Decision Tree Approach." Journal of the Operational Research Society, 61(3), 393-398.

Moral, G. 2006. "EAD Estimates for Facilities with Explicit Limits," in: Engelmann B, Rauhmeier R (Eds), The Basel II Risk Parameters: Estimation, Validation and Stress Testing, Springer, Berlin, 197-242.

Taplin, R., Huong, M. and Hee, J. 2007. "Modeling exposure at default, credit conversion factors and the Basel II Accord." Journal of Credit Risk, 3(2), 75-84.

Valvonis, V. 2008. "Estimating EAD for retail exposures for Basel II purposes." Journal of Credit Risk, 4(1), 79-109.

Chapter 6 Stress Testing

6.1 Overview of Stress Testing

In previous sections, we have detailed how analysts can develop and model each of the different risk parameters that are required under the advanced internal ratings-based (A-IRB) approach. However, this is only part of the journey, as before models can be set free into the wild they must be stress tested over a number of different extrinsic scenarios that can have an impact on them. The topic of stress testing, as defined by the Committee on the Global Financial System (2005),

> *"is a risk management tool used to evaluate the potential impact on a firm of a specific event and/or movement in a set of financial variables. Accordingly, stress-testing is used as an adjunct to statistical models such as value-at-risk (VaR), and increasingly it is viewed as a complement, rather than as a supplement to these statistical measures".*

The importance here of stress testing to financial institutions is in the understanding of how these extrinsic factors, such as the global economy, can directly impact on internal models developed by the financial institutions themselves. Stress testing encapsulates potential impacts of specific situations outside of the control of an organization and acts as a barometer to future events. Typically statistical models, such as Value at Risk, are used to capture risk values with a certain hypothesized probability with stress testing used to consider the very unusual events (such as those 1 in 1000 observed events highlighted in black in the following Figure 6.1).

Figure 6.1: Unusual Events Captured by Stress Tests

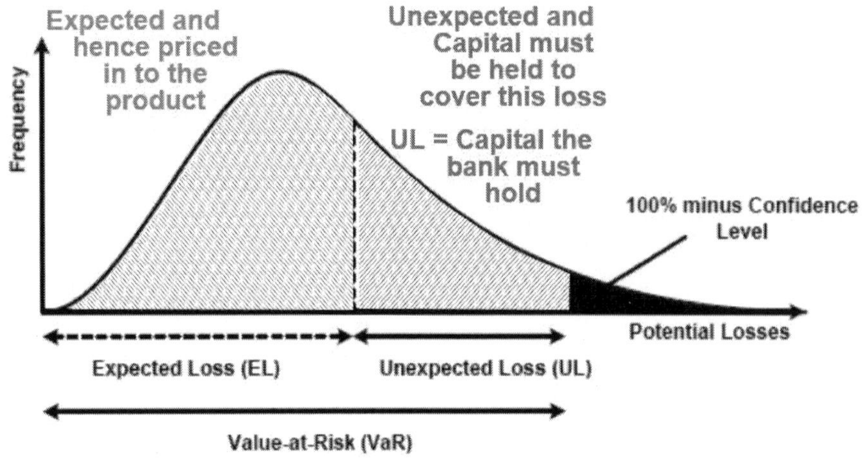

In this chapter, we look at the stress testing techniques and methodologies that can be utilized in the continued support of a model through its lifecycle.

6.2 Purpose of Stress Testing

Financial institutions are under obligation by the regulatory bodies to prove the validity of their models and to show an understanding of the risk profile of the organization. In addition to the regulatory value from conducting stress testing to the financial organization, substantial additional value to the business can also be obtained through this process. For example, in the retail finance sector, forecasting using scenario analysis is often utilized for price-setting. It can also be used in determining which product features should be targeted at which segmented customer group. This type of scenario analysis allows organizations to best price and position credit products to improve their competitive advantage within the market. The ability to capture the impact of extreme but plausible loss events which are not accounted for by Value-at-Risk (VaR) are another important factor of stress testing, allowing organizations to better understand abnormal market conditions. Other typical benefits to organizations employing stress testing activities include:

- A better understanding of capital allocation across portfolios
- Evaluation of threats to the organization and business risks – determining exactly how extrinsic factors can impact the business
- A reduction in the reliance on pro-cyclity –organizations not simply moving with the economic cycle
- Ensuring organizations do not simply hold less capital during an economic boom and more during a downturn

Organizations should consider all of these factors in the development of their internal stress testing strategies. The following section discusses in more depth the methodologies available to organizations and how each one can be utilized to better understand extrinsic effects.

6.3 Stress Testing Methods

Stress testing is a broad term which encompasses a number of methodologies organizations will want to employ in understanding the impact of extrinsic effects. Figure 6.2 depicts the constituent types of stress testing.

Figure 6.2: Stress Testing Methodologies

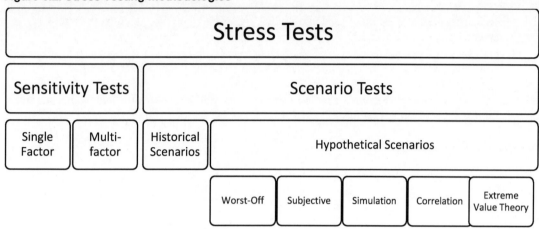

As a concept, stress testing can be broken down into two distinct categories:

1. Sensitivity testing
2. Scenario testing

The following sections detail each of these categories and give examples as to how organizations can implement them in practice.

6.3.1 Sensitivity Testing

Sensitivity testing includes static approaches which do not intrinsically take into account external (macro-economic) information. Typically, these types of stress testing are used for market risk assessment. An example for single factor stress tests is to test how a decrease in the pound to euro exchange rate by 2% will impact on the business. Single factor sensitivity tests for credit risk can be conducted by multiple means:

1. Stressing the data, for example, testing the impact of a decrease in a portfolio of customers' income by 5%;
2. Stressing the PD scores, for example, testing the impact of behavioral scores falling by 15%;
3. Stressing rating grades, for example, testing the impact of an AAA rating grade decreases to AA rating.

The benefit of single factor sensitivity tests are the fact they are relatively easy to implement and understand; however, the disadvantage of this is that they are hard to defend in connection with changes in economic conditions.

Multi-factor sensitivity tests seek to stress all potential factors by understanding the correlation between all of the factors. This type of sensitivity analysis is more synonymous to scenario type testing.

6.3.2 Scenario Testing

Scenario stress testing is the method of taking historical or hypothetical scenario situations where the source of shock is well-defined as well as the parameter values that can be impacted. In most cases, these scenarios are either portfolio or event-driven and can take into account macro-economic factors. Scenario testing can be broken down into two constituent parts:

1. Historical Scenarios –scenarios based upon actual events and therefore potential variations in parameter values are known
2. Hypothetical Scenarios – requires expert judgment to assess potential threats and test against these. Hypothetical scenarios are much harder to conduct as stressing unknown conditions.

Within both of these scenario approaches, there is a trade-off between the reality of what could occur and comprehensibility of the resulting findings.

6.3.2.1 Historical Scenarios

Historical scenario analysis can be used to assess how recent history can impact today's current portfolio position. For example, if a country's exchange rate had experienced a sharp depreciation in the past, this historical scenario could be implemented on a financial institution's current portfolio to assess the impact. In practice, organizations will apply input parameters (such as default rates and collateral values) observed during a downturn year to the current portfolio position.

6.3.2.2 Hypothetical Scenarios

Hypothetical scenarios are applicable when no historic scenario is appropriate, such as when a new portfolio is introduced or new risks are identified. The key to hypothetical scenarios is to take into account all of the potential risk factors and make sure that the combinations of these risk factors make economic sense. Factors that are typically utilized in these hypothetical scenarios include:

- Market downturn – for financial institutions the most common hypothetical scenario is stressing for adverse impacts to the market through macroeconomic factors.
- Market position – loss in competitive position
- Market reputation – experiencing a decline in reputation due to perceived risk

A combination of the above factors should be taken into consideration when developing hypothetical scenario instances.

Categories of Hypothetical Scenarios

There are varying types of hypothetical scenarios that you may wish to consider. These include:

- Correlation scenario analysis: this application utilizes calculated correlation matrices between inputs so that some inputs can be stressed whilst others can be inferred – such as looking at the correlation between PD and LGD values. A downside of this approach is that the same correlations defined in the past may not hold in times of stress.
- Simulation scenario analysis: Simulations can be used to estimate potential outcomes based upon stressed conditions. For example if unemployment were to increase by 2%, 3% or 4%, forecast how this would impact portfolio level loss.
- Worst-case scenario analysis: this type of hypothetical scenario analysis stresses for the most extreme movement in each risk parameter – This is the least plausible and also ignores correlations between risk factors. However, this is also one of the most common approaches.
- Expert judgment scenario analysis: stress testing is conducted on factors determined by expert decisioning – This category of scenario does have a reliance on the knowledge of the experts applying their subjective knowledge.

Stress Testing using Macroeconomic Approaches

It is fundamental to consider macroeconomic factors when applying stress testing approaches in the model development phase. In the financial sector a variety of methodologies are employed to both relate historical defaults to macroeconomic factors and model customer migrations in credit ratings.

Various time series approaches can be applied to analyze the impact of historical macro-economic on past default rates. Typical algorithms utilized for forecasting macro-economic impact include autoregressive integrated moving average (ARIMA) models as well as generalized autoregressive conditional heteroskedasticity (GARCH) models. Both these models can be estimated in SAS using **proc ARIMA** and **proc AUTOREG** respectively.

Cumulative logistic models (**proc logistic** or **Regression node**) are often used in the modeling of changes in credit ratings grades based upon macroeconomic factors.

Frequently used economic variables for this type of analysis are GDP, the unemployment rate, inflation, and house price index (HPI). The assumption is that historical data is readily available, which also includes downturn scenarios. The hypothesis surrounding this does rely heavily on previous recessions being synonymous to future recessions which will not always hold true. The use of a macroeconomic approach can both simulate the effect of historical scenarios as well as hypothetical scenarios.

6.4 Regulatory Stress Testing

In October 2006, a letter was published by the Financial Services Authority (FSA) on a Stress Testing Thematic Review. The key conclusions from this letter focused on:

- "Close engagement by senior management resulted in the most effective stress testing practices."
- "Good practice was observed where firms conducted firm-wide stress tests of scenarios that were plausible, yet under which profitability was seriously challenged, key business planning assumptions were flexed or scope for mitigating action was constrained."
- "Communicating results clearly and accessible are important for many firms."
- "Good practice entailed using group-wide stress testing and scenario analysis to challenge business planning assumptions."

The full FSA Stress Testing Thematic review can be found here: http://www.fsa.gov.uk/pubs/ceo/stress_testing.pdf.

In addition to the FSA report in May 2009, the Basel Committee on Banking Supervision published a report on the "Principles for Sound Stress Testing Practices and Supervision". Within this report, the importance of board and senior management involvement in ensuring the proper design and use of stress testing in bank's risk governance was deemed to be critical. It was shown that for those banks that fared particularly well during the financial crisis, senior management had taken an active interest in the development and operation of strategic stress testing. Account system-wide interactions, feedback effects, and the use of several scenarios should be considered, including forward-looking scenarios. Other findings included:

- Specific risks that should be addressed and are emphasized are securitization risk, pipeline and warehousing risk, reputation risk, and wrong-way risk (part of Pillar 2)
- Importance of supervisory capital assessments and stress testing

The full report can be obtained through the following link http://www.bis.org/publ/bcbs155.pdf.

6.5 Chapter Summary

In summary, this chapter has set out to explain and understand the concepts of stress testing and validation with an aim to inform practitioners of the importance and impact of their use. This section should give readers a greater understanding and appreciation on the topic of stress testing and how through the utilization of the techniques proffered here, models can be managed throughout their lifecycle. In terms of mapping the tests back to the pillars of the Basel Capital Accord (Figure 1.1), sensitivity analysis and static stress tests can be seen as the most appropriate for Pillar 1 testing, whereas scenario analysis and dynamic models are more appropriate for Pillar 2 and capital planning stress testing. Since the 2007/2008 "credit crunch," there has been more and more emphasis put on stress testing by the regulatory bodies as the true cost of those 1 in a 1000 rare events discussed here are realized. It is clear from the 2006 FSA stress testing review that an active involvement of senior levels of management both in defining and also interpreting the output from the stress testing conducted is imperative. A lack of this senior level of buy in will result in ineffective stress testing practices. It can also be summarized that the inclusion of macro-economic drivers is likely to result in an improvement in stress testing practices.

6.6 References and Further Reading

Basel Committee on Banking Supervision. 2009. Principles for sound stress testing practices and supervision, May: http://www.bis.org/publ/bcbs155.pdf

Committee on the Global Financial System. 2005. "Stress Testing at Major Financial Institutions: Survey Results and Practice." January.

UK Financial Services Authority (FSA). 2006. "Stress Testing Thematic Review," letter to chief executives at ten large banking firms. October: http://www.fsa.gov.uk/pubs/ceo/stress_testing.pdf

Chapter 7 Producing Model Reports

7.1 Surfacing Regulatory Reports

In this chapter, we focus on the reporting outputs that are required under regulatory guidelines in the presentation of the internally built PD, LGD, and EAD models. SAS provides a number of ways to produce these reports. Throughout this chapter, the following software is referenced:

1. SAS Enterprise Guide
2. SAS Enterprise Miner
3. SAS Model Manager

We will begin by outlining the types of model validation that are commonly practiced within the industry, alongside the relative performance metrics and statistics that accompany these approaches.

7.2 Model Validation

Under Basel II, there are a number of statistical measures that must be reported on to validate the stability, performance, and calibration of the risk models discussed in this book. The measures utilized in the validation of internal models can be subdivided into three main categories, listed in Table 7.1.

Table 7.1: Model Validation Categories

Category	Description
Model Performance	Measures the ability of a model to discriminate between customers with accounts that have defaulted, and customers with accounts that have not defaulted. The score difference between non-default and default accounts helps to determine the required cutoff score. The cutoff score helps to predict whether a credit exposure is a default account.
	Measures the relationship between the actual default probability and the predicted default probability. This helps you to understand the performance of a model over a time period.
Model Stability	The purpose of this validation approach is to monitor and track changes in the distribution across both the modeling and scoring data sets.

Category	Description
Model Calibration	Model calibration is used to assess the accuracy of PD, LGD, and EAD models and how well these estimated values fit to the data.

The following sections describe the measures, statistics, and tests that are used to create the PD and LGD reports. For more information about these measures, statistics, and tests, see the SAS Model Manager product documentation page on support.sas.com.

7.2.1 Model Performance

As discussed in the previous section, each model validation approach requires a number for performance measures and statistics as part of the reporting process. The following Table 7.2 indicates and briefly defines the performance measures. These reports can be manually generated using SAS Enterprise Miner and SAS/STAT; however, a number of the measures are automatically reported through SAS Model Manager for both PD and LGD (Figure 7.1). For more information on the reports generated in SAS Model Manager, please refer to the following link:

http://support.sas.com/documentation/cdl/en/mdsug/65072/HTML/default/viewer.htm#n194xndt3b3y1pn1ufc0mqbsmht4.htm

Figure 7.1: SAS Model Manager Characteristic and Stability Plots

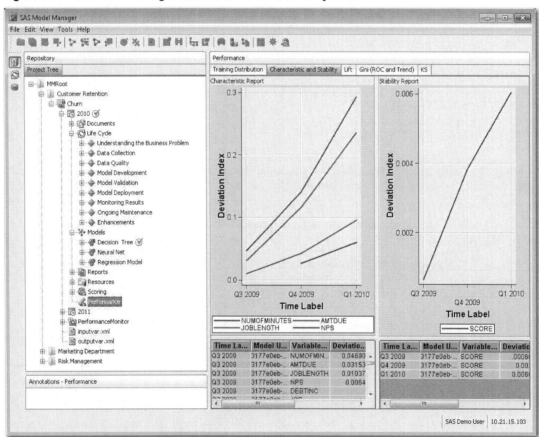

Table 7.2: SAS Model Manager Performance Measures

Performance Measure	Description
Accuracy	Accuracy is the proportion of the total number of predictions that were correct.
Accuracy Ratio (AR)	AR is the summary index of Cumulative Accuracy Profile (CAP) and is also known as Gini Coefficient. It shows the performance of the model that is being evaluated by depicting the percentage of defaulted accounts that are captured by the model across different scores.
Area Under Curve (AUC)	AUC can be interpreted as the average ability of the rating model to accurately classify non-default accounts and default accounts. It represents the discrimination between the two populations. A higher area denotes higher discrimination. When AUC is 0.5, it means that non-default accounts and default accounts are randomly classified, and when AUC is 1, it means that the scoring model accurately classifies non-default accounts and default accounts. Thus, the AUC ranges between 0.5 and 1.
Bayesian Error Rate (BER)	BER is the proportion of the whole sample that is misclassified when the rating system is in optimal use. For a perfect rating model, the BER has a value of zero. A model's BER depends on the probability of default. The lower the BER and the lower the classification error, the better the model.
D Statistic	The D Statistic is the mean difference of scores between default accounts and non-default accounts, weighted by the relative distribution of those scores.
Error Rate	The Error Rate is the proportion of the total number of incorrect predictions.
Information Statistic (I)	The information statistic value is a weighted sum of the difference between conditional default and conditional non-default rates. The higher the value, the more likely it is that a model can predict a default account.
Kendall's Tau-b	Kendall's Tau-b is a nonparametric measure of association based on the number of concordances and discordances in paired observations. Kendall's Tau values range between -1 and +1, with a positive correlation indicating that the ranks of both variables increase together. A negative association indicates that as the rank of one variable increases, the rank of the other variable decreases.
Kullback-Leibler Statistic (KL)	KL is an asymmetrical measure of the difference between the distributions of default accounts and non-default accounts. This score has similar properties to the information value.
Kolmogorov-Smirnov Statistic (KS)	KS is the maximum distance between two population distributions. This statistic helps to discriminate between default accounts and non-default accounts. It is also used to determine the best cutoff in application scoring. The best cutoff maximizes KS, which becomes the best differentiator between the two populations. The KS value can range between 0 and 1, where 1 implies that the model is perfectly accurate in predicting default accounts or separating the two populations. A higher KS denotes a better model.
1–PH Statistic (1–PH)	1-PH is the percentage of cumulative non-default accounts for the cumulative 50% of the default accounts.
Mean Square Error (MSE), Mean Absolute Deviation (MAD), and Mean Absolute Percent Error (MAPE)	MSE, MAD, and MAPE are generated for LGD reports. These statistics measure the differences between the actual LGD and predicted LGD.

Performance Measure	Description
Pietra Index	The Pietra Index is a summary index of Receiver Operating Characteristic (ROC) statistics because the Pietra Index is defined as the maximum area of a triangle that can be inscribed between the ROC curve and the diagonal of the unit square.
Precision	Precision is the proportion of the actual default accounts among the predicted default accounts.
Sensitivity	Sensitivity is the ability to correctly classify default accounts that have actually defaulted.
Somers' D (p-value)	Somers' D is a nonparametric measure of association that is based on the number of concordances and discordances in paired observations. It is an asymmetrical modification of Kendall's Tau. Somers' D differs from Kendall's Tau in that it uses a correction only for pairs that are tied on the independent variable. Values range between -1 and +1. A positive association indicates that the ranks for both variables increase together. A negative association indicates that as the rank of one variable increases, the rank of the other variable decreases.
Specificity	Specificity is the ability to correctly classify non-default accounts that have not defaulted.
Validation Score	The validation score is the average scaled value of seven distance measures, anchored to a scale of 1 to 13, lowest to highest. The seven measures are the mean difference (D), the percentage of cumulative non-default accounts for the cumulative 50% of the default accounts (1-PH), the maximum deviation (KS), the Gini Coefficient (G), the information statistic (I), the Area Under the Curve (AUC), or Receiver Operating Characteristic (ROC) statistic, and the Kullback-Leibler statistic (KL).

An example of the PD accuracy ratio analysis reports created by SAS Model Manager is displayed in Figure 7.2. These reports can be exported as a PDF or RTF pack and/or distributed internally via email.

Figure 7.2: Example PD Accuracy Ratio Analysis

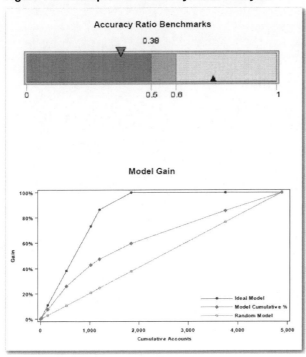

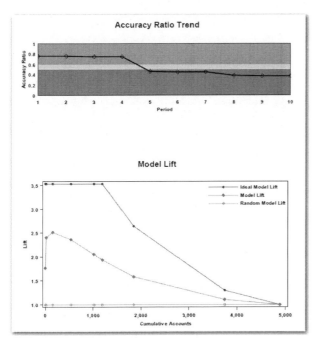

SAS Enterprise Miner provides Model Gain and Model Lift graphics as standard for all of the modeling nodes as well as the **Model Comparison node**. To replicate these graphics in SAS Enterprise Guide, the following steps can be applied:

Step 1 (Program 7.1) creates a Model Gains table from the output of your model which contains the cumulative number of accounts (CUM_ACC), a random (RAND) line value, the cumulative % captured response rate (CAPC) and the best cumulative % captured response rate (BESTCAPC).

Program 7.1: Model Gains Table Code

```
data gain;
input CUM_ACC RAND CAPC BESTCAPC ;
cards;
0 0 0 0
100 0.02 0.08 0.12
500 0.1 0.25 0.39
1000 0.2 0.41 0.75
1250 0.25 0.48 0.85
1750 0.35 0.6 1
3750 0.75 0.83 1
5000 1 1 1 1
;
run;
```

Step 2 (Program 7.2) uses the GPLOT procedure with overlay to plot all three lines against the cumulative number of accounts.

Program 7.2: Plot Procedure Code

```
proc gplot data = GAIN;
plot BESTCAPC * CUM_ACC CAPC * CUM_ACC RAND * CUM_ACC  /
 overlay;
run;
quit;
```

The resulting output submitted in SAS Enterprise Guide can be viewed below in Figure 7.3.

Figure 7.3: Model Gain Chart in SAS Enterprise Guide

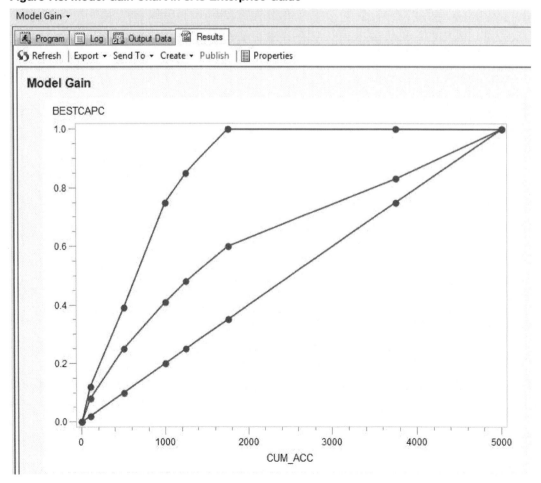

Users can also utilize the **Line Plot...** task in the **Graph** tab of Enterprise Guide using a multiple vertical column line plot with overlay option to achieve the same outcome.

To graphically represent an Accuracy Ratio Trend in SAS Enterprise Guide, the following SGPLOT code with color-coded band statements can be utilized. Step 1 (Program 7.3) defines a table detailing the Accuracy Ratio values (AR) and the time period (Period):

Program 7.3: Accuracy Ratio Table Code

```
data Accuracy_Ratio_Table;
input Period AR;
cards;
1 0.8
2 0.8
3 0.8
4 0.8
5 0.45
6 0.45
7 0.45
8 0.4
9 0.4
10 0.4
;
run;
```

Step 2 (Program 7.4) defines cutoff levels for displaying traffic light warning levels for degradation in model accuracy. Here, we define Bad where the accuracy falls below 0.5, Warning between 0.5 and 0.6, and a Good between 0.6 and 1:

Program 7.4: Cutoff Data Step Code

```
/* Add the "computed" band levels */
data Accuracy_Ratio_Table_Bands;
  set Accuracy_Ratio_Table;
  Bad = 0.5;
  Warning = 0.6;
  Good = 1;
run;
```

Step 3 (Program 7.5) defines the SGPLOT procedural code to process the Accuracy Ratio Table and apply bands for the chart's background. Define the series for the x-axis = Period and the y-axis = Accuracy Ratio.

Program 7.5: Accuracy Ratio Trend Plot Code

```
/* Plot the Accuracy Ratio Trend with bandings */
Title "Accuracy Ratio Trend";
proc sgplot data=Accuracy_Ratio_Table_Bands;
band x=period upper=Good lower=Warning / name="Good" legendlabel="Good"
     fillattrs=(color="green") transparency=0.5;

band x=period upper=Warning lower=Bad / name="Warning" legendlabel="Warning"
     fillattrs=(color="yellow") transparency=0.5;

band x=period upper=Bad lower=0 / name="Bad" legendlabel="Bad"
     fillattrs=(color="red") transparency=0.5;
series x=period y=AR / markers
     lineattrs=(pattern=solid) name="series";
run;
```

The output resulting from a SAS Enterprise Guide program can be viewed below in Figure 7.4:

Figure 7.4: Accuracy Ratio Trend Chart in SAS Enterprise Guide

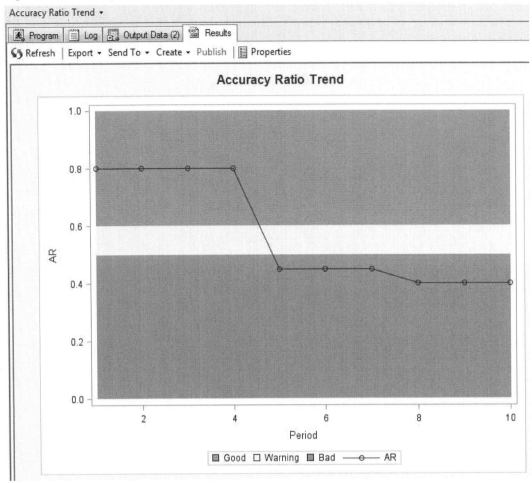

It is advised that users identify their own appropriate bandings based on business requirements. The supplied code can also be utilized for other trend graphics detailed in this chapter, such as the System Stability Index trend in the following section.

7.2.2 Model Stability

Table 7.3 describes the model stability measure that is used to create the PD report and the LGD reports.

Table 7.3: SAS Model Manager Model Stability Performance Measures

Performance Measure	Description
System Stability Index (SSI)	SSI monitors the score distribution over a time period. This can be calculated for both PD and LGD models.

Figure 7.5 displays an example PD stability analysis report generated in the Model Manager environment. These plots display how stable a model's score distribution is across each time period, broken down by data pool.

Figure 7.5: Example PD Stability Analysis

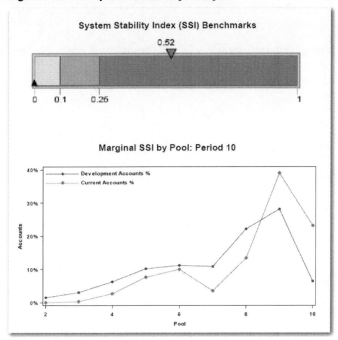

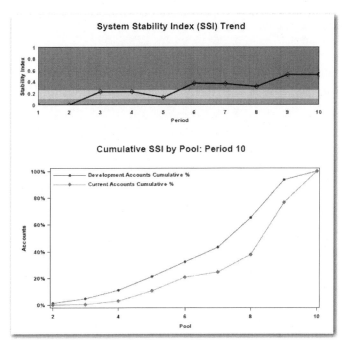

To create a system stability index plot in SAS Enterprise Guide, analysts can utilize the GKPI procedure, which creates graphical key performance indicator (KPI) charts. The following example code shows how a calculated SSI value can be plotted using a horizontal slider chart. Step 1 (Program 7.6) creates a table with the SSI value and the upper boundary of the slider chart:

Program 7.6: SSI Value Table

```
data gkpi;
input SSI Upper;
cards;
0.52 1
;
run;
```

Step 2 (Program 7.7) assigns these values to respective macro variables, where the value for SSI is assigned to mac_actual_1 and Upper is assigned to mac_target_1, for use in the later code:

Program 7.7: Macro Variable Code

```
data actual_1;
set GKPI;
call symput("mac_actual_1",'SSI'n);
call symput("mac_target_1",'Upper'n);
run;
```

Step 3 (Program 7.8) assigns boundary values to macro variables to identify cutoff points (0, 0.1, 0.25, and 1) within our KPI chart:

Program 7.8: Cutoff Code

```
data bounds;
    call symput("mac_b1",0);
    call symput("mac_b2",0.1);
    call symput("mac_b3",0.25);
    call symput("mac_b4",1);
run;
```

Finally Step 4 (Program 7.9) runs the GKPI procedure to create a "slider" chart based on the bounds, actual, and target macro variables assigned in the previous steps:

Program 7.9: SSI Benchmarks GKPI Procedure Code

```
goptions reset=all device=javaimg vsize= 2.50 in hsize= 2.50 in;
proc gkpi mode=basic;
 slider actual=&mac_actual_1 bounds= (&mac_b1 &mac_b2 &mac_b3 &mac_b4) /
 target=&mac_target_1
 label='SSI Benchmarks' lfont=( h= 20 PT c= CX3366FF)
 colors=(cxE6E6E6 cxB2B2B2 cxD06959);
 run;
 quit;
```

After submitting the above code in Enterprise Guide, the resulting plot is shown in Figure 7.6.

Figure 7.6: System Stability Index Plot

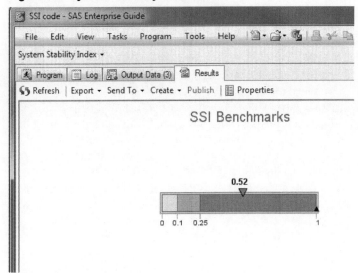

7.2.3 Model Calibration

Table 7.4 describes the model calibration measures and tests that are used to create the PD and LGD reports:

Table 7.4: SAS Model Manager Model Calibration Performance Measures

Performance Measure	Description
Binomial Test	The binomial test evaluates whether the PD of a pool is underestimated. If the number of default accounts per pool exceeds either the low limit (binomial test at 0.95 confidence) or high limit (binomial test at 0.99 confidence), the test suggests that the model is poorly calibrated.
Brier Skill Score (BSS)	BSS measures the accuracy of probability assessments at the account level. It measures the average squared deviation between predicted probabilities for a set of events and their outcomes. Therefore, a lower score represents a higher accuracy.
Confidence Interval (CI)	The CI indicates the confidence interval band of the PD or LGD for a pool. The Probability of Default report compares the actual and estimated PD rates with the CI limit of the estimate. If the estimated PD lies in the CI limits of the actual PD model, the PD performs better in estimating actual outcomes.
Correlation Analysis	The model validation report for LGD provides a correlation analysis of the estimated LGD with the actual LGD. This correlation analysis is an important measure for a model's usefulness. The Pearson correlation coefficients are provided at the pool and overall levels for each time period are examined.
Hosmer-Lemeshow Test (p-value)	The Hosmer-Lemeshow test is a statistical test for goodness-of-fit for classification models. The test assesses whether the observed event rates match the expected event rates in pools. Models for which expected and observed event rates in pools are similar are well-calibrated. The p-value of this test is a measure of the accuracy of the estimated default probabilities. The closer the p-value is to zero, the poorer the calibration of the model.
Mean Absolute Deviation (MAD)	MAD is the distance between the account level estimated and the actual LGD, averaged at the pool level.

Performance Measure	Description
Mean Absolute Percent Error (MAPE)	MAPE is the absolute value of the account level difference between the estimated and the actual LGD, divided by the estimated LGD and averaged at the pool level.
Mean Squared Error (MSE)	MSE is the squared distance between the account level estimated and actual LGD, averaged at the pool level.
Normal Test	The Normal Test compares the normalized difference of predicted and actual default rates per pool with two limits estimated over multiple observation periods. This test measures the pool stability over time. If a majority of the pools lie in the rejection region, to the right of the limits, then the pooling strategy should be revisited.
Observed Versus Estimated Index	The observed versus estimated index is a measure of closeness of the observed and estimated default rates. It measures the model's ability to predict default rates. The closer the index is to zero, the better the model performs in predicting default rates.
Traffic Lights Test	The Traffic Lights Test evaluates whether the PD of a pool is underestimated, but unlike the binomial test, it does not assume that cross-pool performance is statistically independent. If the number of default accounts per pool exceeds either the low limit (Traffic Lights Test at 0.95 confidence) or high limit (Traffic Lights Test at 0.99 confidence), the test suggests that the model is poorly calibrated.

Figure 7.7 displays the typically PD calibration metrics analysis report generated by the Model Manager. The color-coded bandings related to model performance can be adjusted by the user to display acceptable tolerances relative to the portfolio.

Figure 7.7: Example PD Calibration Metrics Analysis

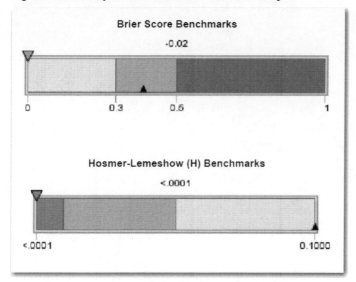

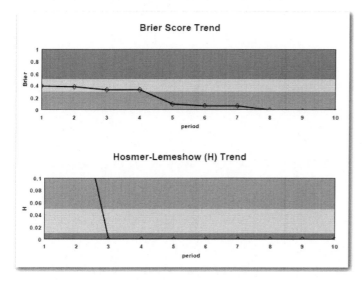

7.3 SAS Model Manager Examples

The following examples demonstrate how users of SAS Model Manager can create and define both PD and LGD reports. The examples shown here can also be found in the SAS Model Manager documentation:

http://support.sas.com/documentation/cdl/en/mdsug/65072/HTML/default/viewer.htm#p0fq2u06em4dc4n10ukj6as24shq.htm

7.3.1 Create a PD Report

To create a PD report, follow these steps:

1. Expand the version folder 🗂.
2. Right-click the **Reports** node and select **Reports ▶ New Report**. The New Report window appears (Figure 7.8).

3. Select **Probability of Default Model Validation Report** from the **Type** box.

Figure 7.8: SAS Model Manager New PD Report

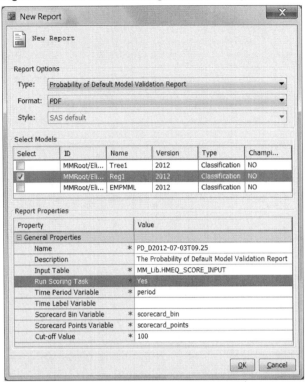

4. In the **Format** box, select the type of output that you want to create. The default is **PDF**. The other option is **RTF**.
5. In the **Select Models** box, select the box for the model for which you want a report.
6. Complete the Report Properties:

   ○ Enter a report name if you do not want to use the default value for the **Name** property.

   ○ Optional: Enter a report description.

   ○ For the **Input Table** property, click the **Browse** button and select a table from the **SAS Metadata Repository** tab or from the **SAS Libraries** tab. The table can contain only input variables, or it can contain input and output variables.

   ○ If the input table contains only input variables, set **Run Scoring Task** to **Yes**. If the input table contains input and output variables, set **Run Scoring Task** to **No**.

   ○ The **Time Period Variable** specifies the variable from the input table whose value is a number that represents the development period. This value is numeric. The time period for PD reports begins with 1. The default is **period**.

   ○ Optional: In the **Time Label Variable** field, enter the variable from the input table that is used for time period labels. When you specify the time label variable, the report appendix shows the mapping of the time period to the time label.

   ○ **Scorecard Bin Variable** is the variable from the input table that contains the scorecard bins. If the scoring job for the PD report is run outside of SAS Model Manager, the scorecard bin variable must be a variable in the input table. If scoring is done within SAS Model Manager, do not include the variable in the input table. The default is **scorecard_bin**.

   ○ **Scorecard Points Variable** is the variable that contains the scorecard points. If the scoring job for the PD report is run outside of SAS Model Manager, the scorecard points variable must be a variable in the input table. If scoring is done within SAS Model Manager, do not include the variable in the input table. The default is **scorecard_points**.

     ○   **Cut-off Value** is the maximum value that can be used to derive the predicted event and to further compute accuracy, sensitivity, specificity, precision, and error rate. The default is **100**.

     Note: The variable names that you specify can be user-defined variables. A variable mapping feature maps the user-defined variables to required variables.

7.   Click **OK**. A dialog box message confirms that the report was created successfully.

7.3.2 Create a LGD Report

To create an LGD report, follow these steps:

1.   Expand the version folder 📁.
2.   Right-click the **Reports** node and select **Reports ▶ New Report**. The New Report window appears (Figure 7.9).
3.   Select Loss Given Default Report from the Type box.

Figure 7.9: SAS Model Manager New LGD Report

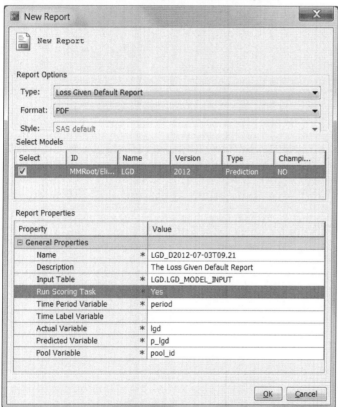

4.   In the **Format** box, select the type of output that you want to create. The default is **PDF**. The other option is **RTF**.
5.   In the **Select Models** box, select the box for the model that you want to report on.
6.   Complete the **Report Properties**. The properties with an asterisk are required.

     ○   Enter a report name if you do not want to use the default value for the **Name** property.

     ○   Optional: Enter a report description.

     ○   For the **Input Table** property, click the **Browse** button and select a table from the **SAS Metadata Repository** tab or from the **SAS Libraries** tab that is used for scoring during the creation of the LGD report. The table can contain only input variables, or it can contain input and output variables.

- ○ If the input table that is specified in the **Input Table** property contains only input variables, set **Run Scoring Task** to **Yes**. If the input table contains input and output variables, set **Run Scoring Task** to **No**.

- ○ The **Time Period Variable** specifies the variable from the input table whose value is a number that represents the development period. This value is numeric. The default is **period**.

- ○ Optional: In the **Time Label Variable** field, enter the variable from the input table that is used for time period labels. When you specify the time label variable, the report appendix shows the mapping of the time period to the time label.

- ○ **Actual Variable** is the actual LGD variable. The default is **LGD**.

- ○ **Predicted Variable** is the project scoring output variable. If the scoring for the LGD report is performed outside of SAS Model Manager, the input data set must include this variable. If the scoring for the LGD report is done by SAS Model Manager, the input data set should not include this variable. The default is **P_LGD**.

- ○ **Pool Variable** is the variable from the input table that is used to identify a two-character pool identifier. The default is **POOL_ID**.

Note: The variable names that you specify can be user-defined variables. A variable mapping feature maps the user-defined variables to required variables.

7. Click **OK**. A dialog box message confirms that the report was created successfully.

7.4 Chapter Summary

In this chapter, we focused on the types of model reports that can be generated from the procedures and methodologies created throughout this book. This chapter will cover the key reporting outputs required within the regulatory framework that can automatically be generated from SAS Model Manager. An example of creating these reports within the SAS Model Manager framework has also been detailed.

Tutorial A – Getting Started with SAS Enterprise Miner

This tutorial shows how to start using SAS Enterprise Miner by demonstrating how to create a new project, define a data library location, import a new data set, create a diagram, and build a simple flow.

SAS Enterprise Miner allows analysts to build models:

- **Faster** – Ease of use through GUI, self-guiding process in SEMMA and a scalable architecture.
- **More Accurately** – Enterprise Miner has a large set of data preparation tools, exploration capabilities, and statistical predictive algorithms that will enable a user to generate the most accurate model that truly represents the relationships and trends within the data.
- **Consistently** – Integration with the SAS platform (data sources, scoring code) and sharable projects that provides a consistent framework.

A.1 Starting SAS Enterprise Miner

Open the SAS Enterprise Miner Client (Figure A.1) and log on with your user details:

Figure A.1: SAS Enterprise Miner Log On

When logged in, the following welcome screen (Figure A.2) will be displayed:

Figure A.2: SAS Enterprise Miner Welcome Screen

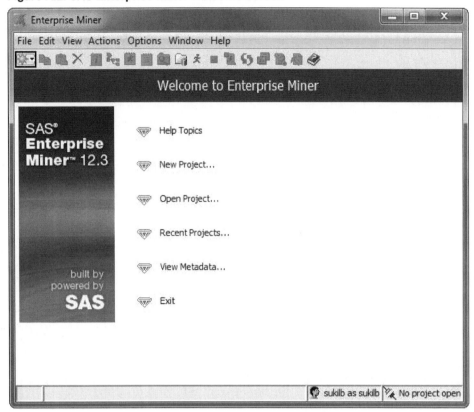

From the welcome screen, click **New Project** and follow the four steps to create a new project. Step 1 will ask you to select a server for the processing of the project. This should have been determined by IT during the installation process. Step 2 will ask you to specify a project name and SAS Server directory, as in the following Figure A.3:

Figure A.3: SAS Enterprise Miner Project Name and Server Directory Screen

Here, we define a new project name as "New Project" and save the directory to our local C:\ drive in a folder called **EM projects**. It may make sense for you to store your project on a shared network directory so that it is backed up regularly and can be accessed by other users on the network. Step 3 will allow you to specify a folder location for registering the project. In this example, we are using the folder location "/User Folders/sasdemo/My Folder". Finally, Step 4 confirms the new project information prior to starting the new project (Figure A.4).

Figure A.4: SAS Enterprise Miner New Project Information Screen

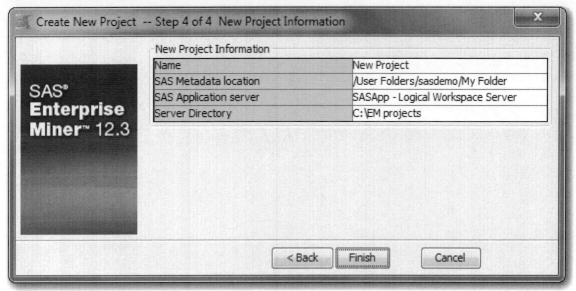

When you click Finish, a new project will be opened (Figure A.5).

Figure A.5: SAS Enterprise Miner New Project Screen

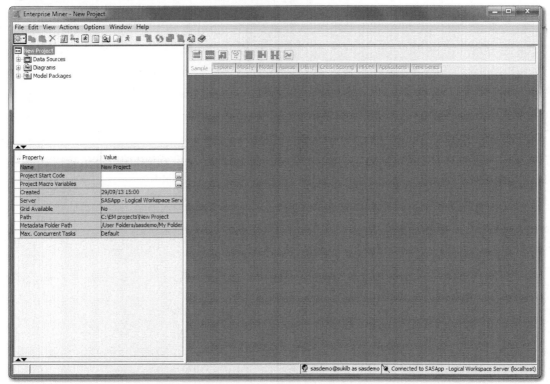

For the purpose of the analysis detailed in this tutorial and referenced throughout this book, you will need to create a library pointing to the SAS data sets accompanying this book. These data sets are available for download from the author's SAS Press page at support.sas.com/authors/ibrown.html. This tutorial will demonstrate the creation of a new process flow within a diagram; however, accompanying XML diagram files have also been provided for reference. The topic of importing a XML diagram is covered in Tutorial B.1.1.

A.2 Assigning a Library Location

A new library location can be defined in two ways:

1. The library wizard
2. Project start up code

To define a library through the library wizard (Figure A.6), click **File ▶ New ▶ Library (or hold Ctrl+Shift+L)**:

Figure A.6: Create a Library

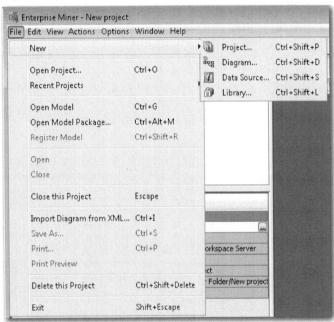

Then follow these steps, illustrated in Figure A.7:

1. Select **Create New Library**, and click **Next**.
2. Specify a name (maximum length of 8 characters). In this example, enter the library name **DemoA**. Specify a library path. In this example, enter the library path **C:\DemoData**.

Figure A.7: Define a Library Name and Location

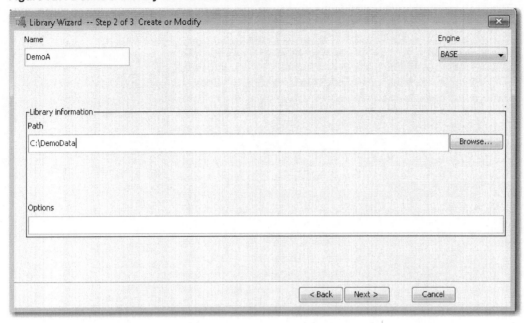

3. Click **Next** to confirm whether the library was successfully defined. A status of "Action Succeeded" will indicate that the library has connected to the file location.

The second approach to defining a new library location in Enterprise Miner is accomplished through writing project start up code.

1. Click the project name, in the following example "New project". You will see an option is available on the Property panel called "Project Start Code" (Figure A.8).

Figure A.8: Create Project Start Code

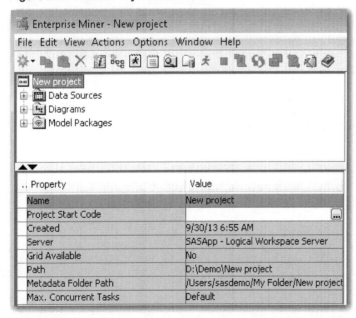

2. Click the ellipsis next to the "Project Start Code" property and the Project Start Code window is displayed. This window behaves in the same way as a traditional editor window in Base SAS or Enterprise Guide, where analysts can write and submit their own code. Within this code editor, submit a libname statement defining the library name as well as the path location (Figure A.9).

3. Click **Run Now** and view the log to confirm that the library has been assigned correctly.

Figure A.9: Write and Submit Project Start Code

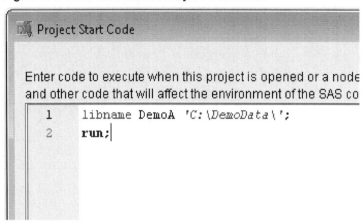

A.3 Defining a New Data Set

Once the library has been correctly defined, you can now create a new data source within Enterprise Miner. For this tutorial, we use the example data accompanying this book.

1. On the Project Panel, click **Data Sources ▶ Create Data Source** (Figure A.10). The source file we want to create is a SAS table, so click **Next** on Step 1.

Figure A.10: Create a New Data Source

2. Browse to the library location (**DemoA**) defined in the previous section. Within this library, select the **KGB** data source (Figure A.11). Click **OK** and then click **Next**.

Figure A.11: Select a SAS Table

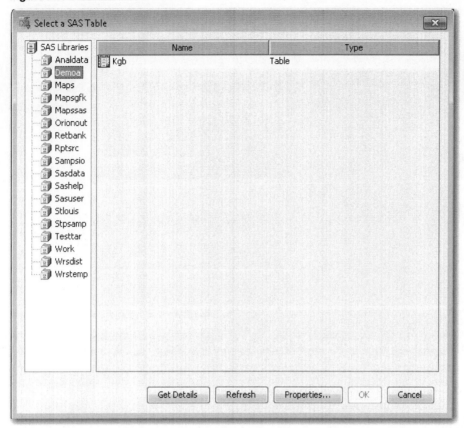

3. Define the table properties to determine that you have selected the right data source based on its size and date created.
4. Click **Next** on Steps 3 and 4.
5. Define how SAS Enterprise Miner will utilize the columns in the data table. By assigning metadata for each of the columns, SAS Enterprise Miner will utilize these columns differently during the model build process. For this tutorial, set the roles and levels for each column as illustrated in Figure A.12.

Figure A.12: Column Roles and Levels

AGE	Input	Interval
BUREAU	Input	Nominal
CAR	Input	Nominal
CARDS	Input	Nominal
CASH	Input	Interval
CHILDREN	Input	Nominal
DIV	Input	Binary
EC_CARD	Input	Binary
FINLOAN	Input	Binary
GB	Target	Binary
INC	Input	Nominal
INC1	Input	Nominal
INCOME	Input	Interval
LOANS	Input	Nominal
LOCATION	Input	Binary
NAT	Input	Nominal
NMBLOAN	Input	Nominal
PERS_H	Input	Nominal
PRODUCT	Input	Nominal
PROF	Input	Nominal
REGN	Input	Nominal
RESID	Input	Binary
STATUS	Input	Nominal
TEL	Input	Nominal
TITLE	Input	Binary
TMADD	Input	Interval
TMJOB1	Input	Interval
freq	Frequency	Binary

For this analysis, all of the variables can be used as inputs except for the variable **GB,** which has the Role of Target and a level of Binary, and **_freq_**, which has a Role of Frequency and a level of Binary.

6. Click **Next** on Step 5 and 6.

7. Assign the role of KGB to **Raw**. By assigning a particular role to the data, Enterprise Miner will only utilize the data set in a particular way. The available options are:

   ○ **Raw** – This is the default setting and enables you to perform your own partitioning into Train, Validation and Test. (We will be using the **Data Partition node** in Tutorial B to split this data before modelling is undertaken).

   ○ **Train** – The data will only be used for model development purposes.

   ○ **Validate** – The data will only be used to tune and validate a model developed on a Train set.

   ○ **Test** – The data will be used as an additional holdout set for model assessment.

   ○ **Score** – The data will only be used to apply scoring logic to.

   ○ **Transaction** – Utilized for longitudinal data.

8. The final step summarizes the data and metadata defined within the data source wizard. When you click **Finish**, the new data source will become active within the data sources tab, as shown in Figure A.13.

Figure A.13: KGB Data Source

Tutorial B – Developing an Application Scorecard Model in SAS Enterprise Miner

In this tutorial, we demonstrate how SAS Enterprise Miner can be utilized to develop a full application scorecard from data sampling to model reporting. We utilize the credit scoring tool set to build the application scorecard and identify whether a customer is likely to be good (paying) or bad (non-paying).

This tutorial details the steps to be taken to demonstrate the various elements of a typical application scorecard development process. This tutorial can be used to explore all of the steps, or individual areas of application scoring, depending on the interest of the analyst.

B.1 Overview

Credit Scoring, as with other predictive models, is a tool used to evaluate the level of risk associated with applicants or customers. It provides statistical odds, or probability, that an applicant with any given score will be "good" or "bad". These probabilities or scores, along with other business considerations such as expected approval rates, profit, churn, and losses, are then used as a basis for decision-making.

This tutorial runs through an example credit scoring model analysis on real-world financial data. The Known Good Bad (KGB) data contains customer information such as age, residential status, and income as well as financial product information such as the type of credit card held by customer and the number of loans. A full credit scoring process with the development of a scorecard is shown through the implementation of this data set.

In this tutorial, we would like to develop a credit scorecard to evaluate the level of risk associated with applicants or customers. SAS Enterprise Miner streamlines the entire credit scoring process from data access to model deployment by supporting all necessary tasks within a single, integrated solution.

By using a credit scorecard to inform our decision-making process, we are more likely to accept potentially good customers and reject potentially bad customers. Therefore, this will reduce the risk of our overall portfolio.

We begin by importing an XML diagram supplied on the author's page, which contains a pre-developed application scorecard process flow. We will then go through the typical steps an analyst would need to cover in running, assessing, and modifying a model.

B.1.1 Step 1 – Import the XML Diagram

In the "New Project" created in Tutorial A, right click the **Diagrams** tab and **Import** the accompanying "Example Credit Scorecard" from XML, as shown in Figure B.1.

Figure B.1: SAS Enterprise Miner Import Diagram Screen

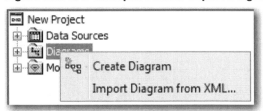

Once imported, the diagram shown in Figure B.2 should be present within your session.

Figure B.2: SAS Enterprise Miner Diagram

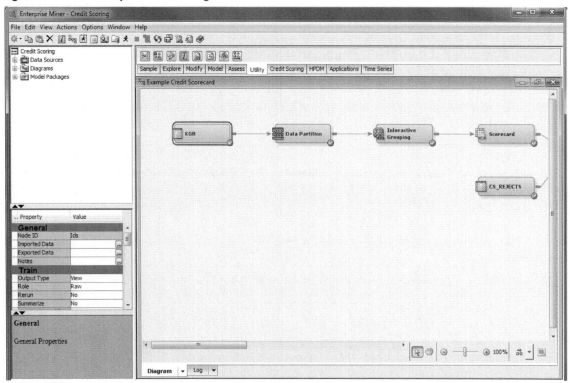

This predefined process flow has been provided for the purpose of this tutorial. This tutorial walks through each of the processes that may be applied in a typical application scorecard project.

B.1.2 Step 2 – Define the Data Source

To highlight and explore data in the KGB data set, the first step an analyst needs to do is bring the data into the EM workbench (see Tutorial A). This data source ideally would have been produced by a centralized team of data architects who have collated data from various sources as part of a data integration process. The data set consists of just the applicants who have been previously extended credit. The data set contains examples of non-event and event customers and is used to train the initial model. The KGB data set contains a binary target variable where 0 indicates records that represent non-event (Good) applicants and 1 indicates records that represent event (Bad) applicants.

Analysts can typically spend about 70% of their time preparing the data for a data mining process. If this data preparation task can be centralized within the IT department, then all analysts are able to get a consistent and clean view of the data that will then allow them to spend the majority of their time building accurate models.

B.1.3 Step 3 – Visualize the Data

Select the KGB data source node in the diagram (Figure B.3) and click **the Property Panel,** found in the ellipsis next to Variables under the **Train – Columns** property panel.

Figure B.3: KGB Data Node

As illustrated in Figure B.4, select **AGE, CAR, CARDS, GB, INCOME** and **RESID,** and click **Explore...**

Figure B.4: Select Variables

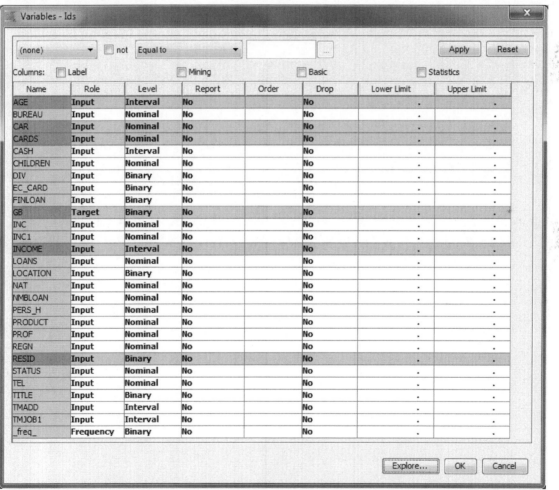

Name	Role	Level	Report	Order	Drop	Lower Limit	Upper Limit
AGE	Input	Interval	No		No	.	.
BUREAU	Input	Nominal	No		No	.	.
CAR	Input	Nominal	No		No	.	.
CARDS	Input	Nominal	No		No	.	.
CASH	Input	Interval	No		No	.	.
CHILDREN	Input	Nominal	No		No	.	.
DIV	Input	Binary	No		No	.	.
EC_CARD	Input	Binary	No		No	.	.
FINLOAN	Input	Binary	No		No	.	.
GB	Target	Binary	No		No	.	.
INC	Input	Nominal	No		No	.	.
INC1	Input	Nominal	No		No	.	.
INCOME	Input	Interval	No		No	.	.
LOANS	Input	Nominal	No		No	.	.
LOCATION	Input	Binary	No		No	.	.
NAT	Input	Nominal	No		No	.	.
NMBLOAN	Input	Nominal	No		No	.	.
PERS_H	Input	Nominal	No		No	.	.
PRODUCT	Input	Nominal	No		No	.	.
PROF	Input	Nominal	No		No	.	.
REGN	Input	Nominal	No		No	.	.
RESID	Input	Binary	No		No	.	.
STATUS	Input	Nominal	No		No	.	.
TEL	Input	Nominal	No		No	.	.
TITLE	Input	Binary	No		No	.	.
TMADD	Input	Interval	No		No	.	.
TMJOB1	Input	Interval	No		No	.	.
freq	Frequency	Binary	No		No	.	.

Figure B.5: Display Variable Interactions

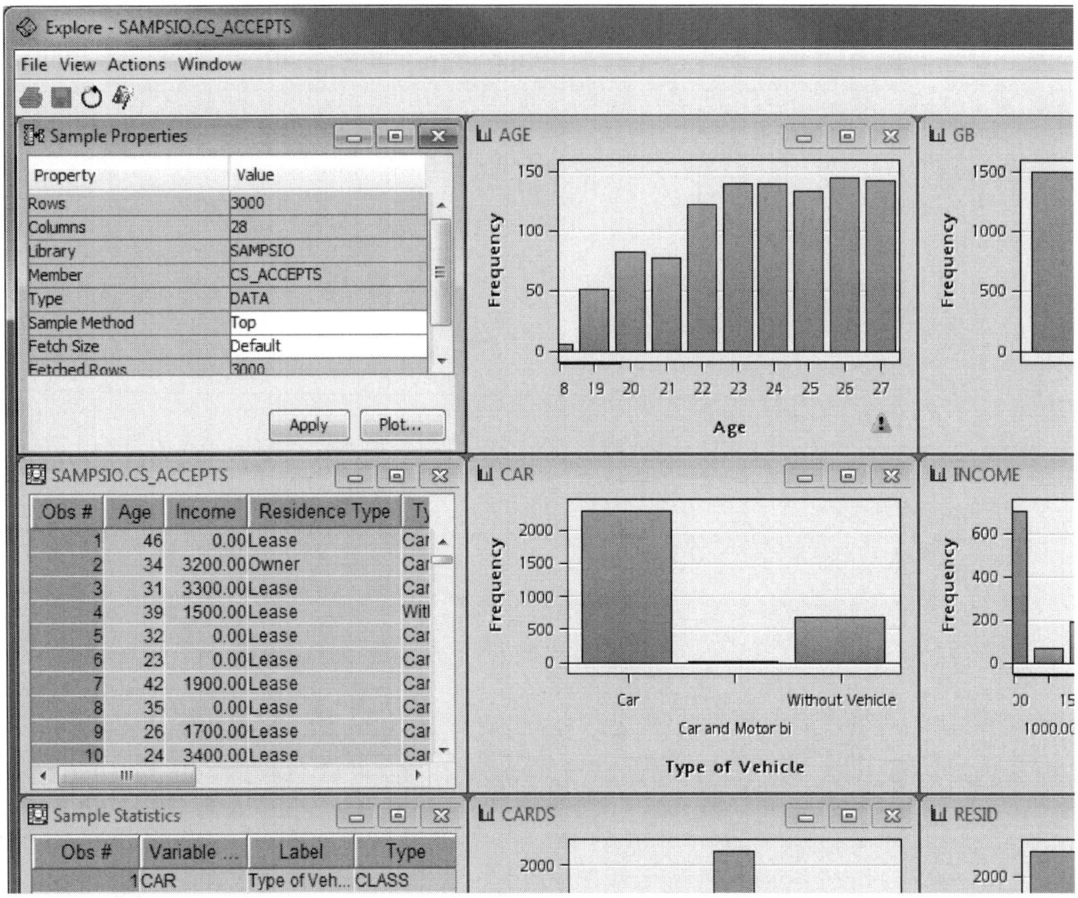

Analysts like to get a full understanding of their data. A quick way to do this is through the interactive data visualization. If you click the **Without Vehicle** bar in the CAR graphic (Figure B.5), you can select all the customers who do not have a car in the other graphics. Right-click a graph, select **Data Options**, and use the **Where** tab to query data displayed.

By using visual interactive data exploration, analysts can quickly assess the quality of the data and any initial patterns that exist. They can then use this to help drive the rest of the data mining project in terms of the modification of the data before they look at the modeling process.

B.1.4 Step 4 – Partition the Data

Figure B.6: Property Panel for the Data Partition Node

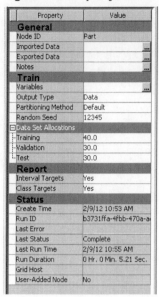

In the process of developing a scorecard, you perform predictive modeling. Thus, it is advisable to partition your data set into training and validation samples. If the total sample size permits, having a separate test sample permits a more robust evaluation of the resulting scorecard. The **Data Partition node**, the property panel for which is shown above in Figure B.6, is used to partition the KGB data set.

B.1.5 Step 5 –Perform Screening and Grouping with Interactive Grouping

To perform univariate characteristic screening and grouping, an **Interactive Grouping node** (Figure B.7) is used in the process flow diagram.

Figure B.7: Interactive Grouping Node

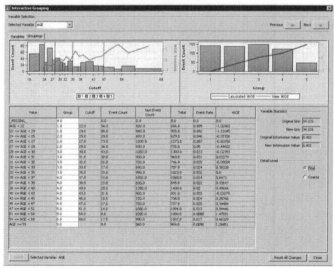

The **Interactive Grouping node** can automatically group the characteristics for you; however, the **Interactive Grouping node** also enables you to perform the grouping interactively (on the Property Panel, select **Train ▶ Interactive Grouping**).

Performing interactive grouping is important because the results of the grouping affect the predictive power of the characteristics, and the results of the screening often indicate the need for regrouping. Thus, the process of grouping and screening is iterative, rather than a sequential set of discrete steps.

Grouping refers to the process of purposefully censoring your data. Grouping offers the following advantages:

- It offers an easier way to deal with rare classes and outliers with interval variables.
- It makes it easy to understand relationships, and therefore gain far more knowledge of the portfolio.
- Nonlinear dependencies can be modeled with linear models.
- It gives the user control over the development process. By shaping the groups, you shape the final composition of the scorecard.
- The process of grouping characteristics enables the user to develop insights into the behavior of risk predictors and to increase knowledge of the portfolio, which can help in developing better strategies for portfolio management.

B.1.6 Step 6 – Create a Scorecard and Fit a Logistic Regression Model

The **Scorecard node** (Figure B.8) fits a logistic regression model and computes the scorecard points for each attribute. With the SAS EM Scorecard you can use either the Weights of Evidence (WOE) variables or the group variables that are exported by the **Interactive Grouping node** as inputs for the logistic regression model.

Figure B.8: Scorecard Node

The **Scorecard node** provides four methods of model selection and seven selection criteria for the logistic regression model. The scorecard points of each attribute are based on the coefficients of the logistic regression model. The **Scorecard node** also enables you to manually assign scorecard points to attributes. The scaling of the scorecard points is also controlled by the three scaling options within the properties of the **Scorecard node**.

B.1.7 Step 7 – Create a Rejected Data Source

The REJECTS data set contains records that represent previous applicants who were denied credit. The REJECTS data set does not have a target variable.

The **Reject Inference node** automatically creates the target variable for the REJECTS data when it creates the augmented data set. The REJECTS data set must include the same characteristics as the KGB data. A role of SCORE is assigned to the REJECTS data source.

B.1.8 Step 8 – Perform Reject Inference and Create an Augmented Data Set

Credit scoring models are built with a fundamental bias (selection bias). The sample data that is used to develop a credit scoring model is structurally different from the "through-the-door" population to which the credit scoring model is applied. The non-event or event target variable that is created for the credit scoring model is based on the records of applicants who were all accepted for credit. However, the population to which the credit scoring model is applied includes applicants who would have been rejected under the scoring rules that were used to generate the initial model. One remedy for this selection bias is to use reject inference. The reject inference approach uses the model that was trained using the accepted applications to score the rejected applications. The observations in the rejected data set are classified as inferred non-event and inferred event. The inferred observations are then added to the KGB data set to form an augmented data set.

This augmented data set, which represents the "through-the-door" population, serves as the training data set for a second scorecard model.

SAS EM provides the functionality to conduct three types of reject inference:

- **Fuzzy**—Fuzzy classification uses partial classifications of "good" and "bad" to classify the rejects in the augmented data set. Instead of classifying observations as "good" and "bad," fuzzy classification allocates weight to observations in the augmented data set. The weight reflects the observation's tendency to be good or bad. The partial classification information is based on the p(good) and p(bad) from the model built on the KGB for the REJECTS data set. Fuzzy classification multiplies the p(good) and p(bad) values that are calculated in the Accepts for the Rejects model by the user-specified Reject Rate parameter to form frequency variables. This results in two observations for each observation in the Rejects data. One observation has a frequency variable (Reject Rate * p(good)) and a target variable of 0, and the other has a frequency variable (Reject Rate * p(bad)) and a target value of 1. Fuzzy is the default inference method.

- **Hard Cutoff**—Hard Cutoff classification classifies observations as "good" or "bad" observations based on a cutoff score. If you choose Hard Cutoff as your inference method, you must specify a Cutoff Score in the Hard Cutoff properties. Any score below the hard cutoff value is allocated a status of "bad." You must also specify the Rejection Rate in General properties. The Rejection Rate is applied to the REJECTS data set as a frequency variable.

- **Parceling**—Parceling distributes binned scored rejects into "good" and bad" based on expected bad rates p(bad) that are calculated from the scores from the logistic regression model. The parameters that must be defined for parceling vary according to the Score Range method that you select in the Parceling Settings section. All parceling classifications, as well as bucketing, score range, and event rate increase, require the Reject Rate setting.

B.1.9 Step 9 – Partition the Augmented Data Set into Training, Test and Validation Samples

The augmented data set that is exported by the **Reject Inference node** is used to train a second scorecard model. Before training a model on the augmented data set, a second data partition is included in the process flow diagram, which partitions the augmented data set into training, validation, and test data sets.

B.1.10 Step 10 – Perform Univariate Characteristic Screening and Grouping on the Augmented Data Set

As we have altered the sample by the addition of the scored rejects data, a second **Interactive Grouping node** is required to recompute the weights of evidence, information values, and Gini statistics. The event rates have changed, so regrouping the characteristics could be beneficial.

B.1.11 Step 11 – Fit a Logistic Regression Model and Score the Augmented Data Set

The final stage in the credit scorecard development is to fit a logistic regression on the augmented data set and to generate a scorecard (an example of which is shown in Figure B.9) that is appropriate for the "through-the-door" population of applicants.

Figure B.9: Example Scorecard Output

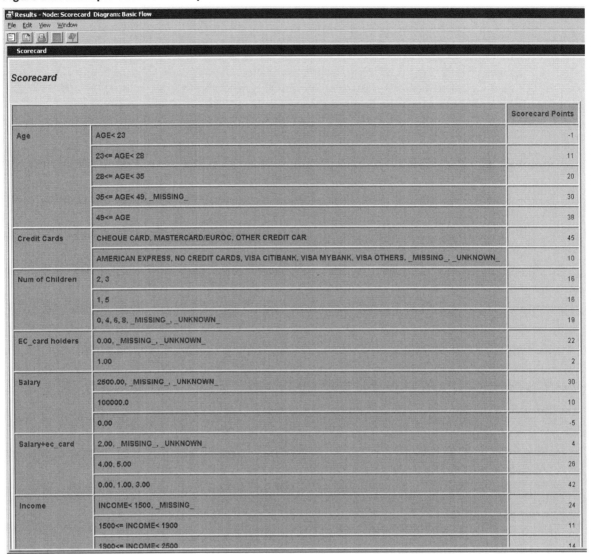

Right-click the **Scorecard node** and select **Results...**, then maximize the Scorecard tab to display the final scores assigned to each characteristic.

B.2 Tutorial Summary

We have seen how the credit scoring nodes in SAS Enterprise Miner allow an analyst to quickly and easily create a credit scoring model using the functionality of the **Interactive Grouping node**, **Reject Inference node,** and **Scorecard node** to understand the probability of a customer being a good or bad credit risk.

Appendix A Data Used in This Book

A.1 Data Used in This Book

Throughout this book, a number of data sets have been utilized in demonstration of the concepts discussed. To enhance the reader's experience, go to support.sas.com/authors and select the author's name to download the accompanying data tables. Under the title of this book, select **Example Code and Data** and follow the instructions to download the data.

The following information details the contents of each of the data tables and the chapter in which each has been used.

Chapter 3: Known Good Bad Data

Filename: KGB.sas7bdat

File Type: SAS Data Set

Number of Variables: 28

Number of Observations: 3000

Variables:

Name	Label	Role	Level
AGE	Age	Input	Interval
BUREAU	Credit Bureau Risk Class	Input	Nominal
CAR	Type of Vehicle	Input	Nominal
CARDS	Credit Cards	Input	Nominal
CASH	Requested cash	Input	Interval
CHILDREN	Num of Children	Input	Nominal
DIV	Large region	Input	Binary
EC_CARD	EC_card holders	Input	Binary
FINLOAN	Num finished Loans	Input	Binary
GB	Good/Bad	Target	Binary
INC	Salary	Input	Nominal
INC1	Salary+ec_card	Input	Nominal
INCOME	Income	Input	Interval
LOANS	Num of running loans	Input	Nominal
LOCATION	Location of Credit Bureau	Input	Binary
NAT	Nationality	Input	Nominal
NMBLOAN	Num Mybank Loans	Input	Nominal
PERS_H	Num in Household	Input	Nominal
PRODUCT	Type of Business	Input	Nominal
PROF	Profession	Input	Nominal
REGN	Region	Input	Nominal
RESID	Residence Type	Input	Binary
STATUS	Status	Input	Nominal
TEL	Telephone	Input	Nominal
TITLE	Title	Input	Binary
TMADD	Time at Address	Input	Interval
TMJOB1	Time at Job	Input	Interval
freq		Frequency	Binary

Chapter 3: Rejected Candidates Data

Filename: REJECTS.sas7bdat

File Type: SAS Data Set

Number of Variables: 26

Number of Observations: 1,500

Variables: Contains the same information as the KGB data set, minus the GB target flag and _freq_ flag.

Chapter 4: LGD Data

Filename: LGD_Data.sas7bdat

File Type: SAS Data Set

Number of Variables: 15

Number of Observations: 3000

Variables:

Name	Role	Level
Account_ID	ID	Nominal
Input1	Input	Interval
Input10	Input	Interval
Input11	Input	Interval
Input12	Input	Interval
Input13	Input	Interval
Input2	Input	Interval
Input3	Input	Interval
Input4	Input	Interval
Input5	Input	Interval
Input6	Input	Interval
Input7	Input	Interval
Input8	Input	Interval
Input9	Input	Interval
LGD	Target	Interval

Chapter 5: Exposure at Default Data

Filename: CCF_ABT.sas7bdat

File Type: SAS Data Set

Number of Columns: 11

Number of Observations: 3,082

Variables:

Name	Role	Level
Cohort	Segment	Binary
Commitment_Size	Input	Interval
Credit_Percentage	Input	Interval
Drawn	Input	Interval
EAD	Target	Interval
ID	ID	Nominal
Rating_Class	Input	Nominal
Time_to_Default	Input	Nominal
Undrawn	Input	Interval
ccf	Target	Interval
prodtype	Input	Nominal

Index

A

Accuracy performance measure 117
Accuracy Ratio (AR) performance measure 54, 117
Accuracy Ratio Trend, graphically representing in SAS
        Enterprise Guide 121–122
advanced internal ratings-based approach (A-IRB) 2
Analytical Base Table (ABT) format 50
application scorecards
    about 35
    creating 144
    data partitioning for 40
    data preparation for 37–38
    data sampling for 39–40
    developing models in SAS Enterprise Miner 139–
        145
    developing PD model for 36–47
    filtering for 40
    input variables for 37–38
    for Known Good Bad Data (KGB) 39
    model creation process flow for 38
    model validation for 46–47
    modeling for 41–45
    motivation for 36–37
    outlier detection for 40
    reject inference for 45–46
    scaling for 41–45
    strength of 54
    transforming input variables for 40–41
    variable classing and selection for 41
application scoring 16
Area Over the Curve (AOC) 71
Area Over the Regression Error Characteristic (REC)
        Curves 71–72
Area Under Curve (AUC) 54, 70–72, 117
ARIMA procedure 113
Artificial Neural Networks (ANN) 63, 67, 79
assigning library locations 134–136
augmented data sets
    creating 144–145
    grouping 145
    partitioning into training, test and validation 145
    scoring 145
augmented good bad (AGB) data set 46
AUOTREG procedure 113

B

Basel Committee on Banking Supervision 4, 8
Basel II Capital Accord 2, 4
Basel III 3
Bayesian Error Rate (BER), as performance measure
        117
behavioral scoring
    about 17, 47

    data preparation for 49–50
    developing PD model for 49–52
    input variables for 49
    model creation process flow for 50–52
    motivation for 48
benchmarking algorithms for LGD 77–82
Beta Regression (BR) 63, 65–67
beta transformation, linear regression nodes combined
        with 65
Binary Logit models 98–99
binary variables 15
Binomial Test 125
"black-box" techniques 44
Box-Cox transformation, linear regression nodes
        combined with 63
Brier Skill Score (BSS) 125

C

calibration, of Probability of Default (PD) models 29
capital requirement (K) 6
Captured Event Plot 54
case study: benchmarking algorithms for LGD 77–82
classification techniques, for Probability of Default
        (PD) models 29–35
Cluster node (SAS Enterprise Miner) 24–25
Cohort Approach 89
Confidence Interval (CI) 125
corporate credit
    Loss Given Default (LGD) models for 60–61
    Probability of Default (PD) models for 28
Correlation Analysis 125
correlation factor (R) 7
correlation scenario analysis 112
creating
    application scorecards 144
    augmented data sets 144–145
    Fit Logistic Regression Model 145-146
    Loss Given Default (LGD) reports 129–130
    Probability of Default (PD) reports 127–129
    rejected data source 144
creation process flow
    application scorecards 39
    for behavioral scoring 50–52
    for Loss Given Default (LGD) 74–75
credit conversion factor (CCF)
    about 92
    distribution 93–94
    time horizons for 88–90
credit risk modeling 2–3
Cumulative Logit models 30, 98–99
cumulative probability 30

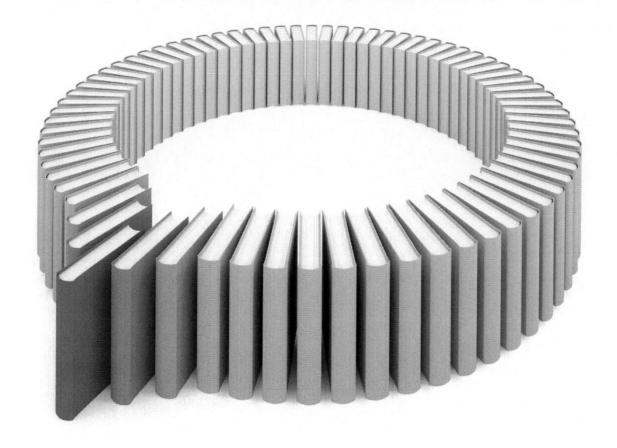

Gain Greater Insight into Your SAS® Software with SAS Books.

Discover all that you need on your journey to knowledge and empowerment.

 support.sas.com/bookstore
for additional books and resources.

THE POWER TO KNOW®